ON April 19, 1956, Commander Lionel "Buster" Crabb, O.B.E., G.M., R.N.V.R., ace frogman and one of the world's greatest experts on underwater tactics and warfare, mysteriously disappeared. He had dived into Portsmouth Harbor to examine the hull of the Soviet cruiser *Ordzhonikidze* . . . he never returned. After ten days, the Admiralty announced he must be presumed dead. Fourteen months later—after three Soviet submarines passed through the English Channel—a headless and handless body was officially identified as the remains of Commander Crabb and buried at Milton Cemetery, Portsmouth, England.

But Crabb did not die. Documents secured by the anti-Communist underground proved that he was captured by the Russians, drugged, and taken to Moscow, escorted by six MVD officers, where he fell victim to the latest methods of Soviet brainwashing.

Commander Crabb Is Alive is the fully documented and authenticated account of what really happened to the missing frogman, bringing the story of Crabb's life behind the Iron Curtain right up to date with details of his promotion to captain in the Soviet Navy!

"The author's tenacity to root out the facets of Commander Crabb's mysterious disappearance and to throw light on to the workings of the Soviet Secret Service department which employs killers, saboteurs and kidnappers deserves the highest praise. This true story of detection, high adventure, and sudden violence is not a lurid spy thriller—it is fact."

—Sydney James Knowles, B.E.M.

Commander Crabb Is Alive

J. BERNARD HUTTON

AWARD BOOKS
NEW YORK

TANDEM BOOKS
LONDON

To
Thomas Edward

First published
by Library 33 Ltd.
in Great Britain

First Award printing 1968

AWARD BOOKS are Published by
Universal Publishing and Distributing Corporation
235 East 45th Street, New York, N. Y. 10017

TANDEM BOOKS are Published by
Universal-Tandem Publishing Company Ltd.
33 Beauchamp Place, London SW 3, England

Manufactured in the United States of America

CONTENTS

	Introduction by Cdr. J. S. Kerans, D.S.O.	9
	Preface by S. J. Knowles, B.E.M.	10
	Prologue	13
1.	Journey to Samara	20
2.	The Trap is Sprung	28
3.	Missing . . . Presumed Dead	38
4.	Breakfast with Zhabotin	46
5.	Trouble in the House	56
6.	Point of No Return	64
7.	A Message for Patricia	73
8.	Question of Identity	83
9.	Stranger in a Train	93
10.	The Smuggled Dossier	100
11.	Sound and Fury	107
12.	Third Officer Smith	119
13.	Icebreaker in Difficulties	130
14.	The Woman from Turkey	143
15.	Riddle of a Skull	154
16.	Clamp-down at Boltenhagen	162
17.	Naval Fable	180
	Epilogue	186
	Index	190

Acknowledgments

I wish to express my thanks to—

Cdr. J. P. A. Barr, R.N. (retd.) and his wife Freddie —Commander Crabb's close friends;

Len Cole, Commander Crabb's very close and perhaps oldest friend;

Margaret Elaine Crabb, Commander Crabb's former wife;

Bernard Francis Castle Floud, M.P.;

George Rodney (Bob) French and Johnny Hunt-Crowley;

Eileen Handley and Catherine S. Jarvis, Commander Crabb's close relatives;

Commander J. S. Kerans, D.S.O., R.N. (retd.);

Sydney James Knowles, B.E.M., Commander Crabb's fellow officer and very close friend:

William Robinson, Commander Crabb's diving companion and friend; as well as all his other fellow officers and friends whose names cannot be disclosed because they are on active service in the Royal Navy;

Patricia Rose, Commander Crabb's fiancée;

Sir Percy Sillitoe, K.B.E., C.B.E., former director of Britain's M.I.5;

Jack Thomas, who helped me with this book;

Arthur Ralph Thorpe, B.E.M., Commander Crabb's chief diving instructor during the last war and treasured friend.

Without their invaluable assistance I would have been unable to write this book and unravel the twelve years old mystery.

INTRODUCTION

by:

Commander J. S. Kerans, D.S.O., R.N. (retd.)

Hero of the Amethyst's escape from the Yangtse

This is an interesting book on the Commander Crabb affair which delves deeper into matters than heretofore and presents much new material. Nevertheless, I do not think that the world will ever be presented with a satisfactory 'official' answer to all the items that Mr. Hutton has so ably pieced together, for I do not presume that the Russians are likely to come 'clean' in the accepted sense of the word.

Considering, however, that the Philby memoirs are slowly coming out in pieces in the Western Press, it would not surprise me if the Russians eventually revise their policy of stubborn denial and utilise the Commander Crabb case for their own propaganda—if and when it should suit them. If it is remembered that Nikita Khrushchev first officially declared that he had never heard of Burgess and Maclean and emphatically denied that they were in the Soviet Union but that later the two British defectors were presented at a Moscow 'Press Conference', the possibility that something similar could happen with Commander Crabb is not too far-fetched.

I have examined the complete evidence very carefully but I hesitate to believe conclusively that the body buried at Portsmouth is in fact Commander Crabb's. Even though I am confident that Commander Crabb is or was living in Russia, I wish to make it clear that I have no intention of trying to influence anyone, but that the reader must form his own conclusions.

PREFACE

by:

Sydney James Knowles, B.E.M.

Commander Crabb's closest diving companion and friend

'Buster' Crabb, 'Crabbie,' Commander Lionel Kenneth Philip Crabb, O.B.E., G.M., R.N.V.R., is a name which to millions who love a hero immediately brings to memory the most mysterious and greatest factual espionage story of all time.

His true métier was under the waves of the world's oceans, tackling any dangerous project that the Admiralty, or others, asked him to do. He was blind with recklessness, but very, very skilful in his task—he was not unlike a Knight of the past, with a blind loyalty for his King and Queen that inspired him to overcome all obstacles.

This, then, was the hero who disappeared mysteriously beneath the waves of Portsmouth Harbour on 19 April 1956, whilst attempting to examine the bottoms of the Russian warships which had brought Marshal Bulganin and Mr. Khrushchev to England on a visit of so-called goodwill.

He was so very bitter and unhappy when, in 1948, as a member of the Royal Navy Voluntary Reserves he found that his services were no longer required. 'They have slung me out, Syd,' he complained bitterly to me. 'What *must* it take to make them realise that we haven't yet scratched the surface in the field of aquatic sabotage and anti-sabotage? . . . They don't need me any more!' he exclaimed. But how wrong he was, for, every time a very tricky underwater job came up, the cry was: 'Send for "Buster" Crabb.' He did magnificent underwater work during the loss of H.M. Submarines *Truculent* and *Affray*.

Civilian life was alien to him—he hated it. Deprived of his naval colleagues and career, he endeavoured to live in the past when together we shared countless adventures. He talked of suicide and exclaimed: 'What a bloody state I am in! I must go and have a talk with "God's Butler" '—his name for the priest at Brompton Oratory in London. Thus this unhappy hero single-handed took on the might of the Soviet Navy to make us all—*sit up*.

'Commander Crabb is dead,' said the Admiralty—an accident, whilst working on a secret naval mission in the vicinity of Portsmouth Harbour. He was such dismissed by the Navy.

In one of the most extraordinary announcements ever made by a Prime Minister, Sir Anthony Eden made a short statement concerning Commander Crabb's death, in Parliament on 9 May 1956, when he said *it was not within the public interest to disclose the circumstances of how Commander Crabb met his death and that what had been done was without the authority or knowledge of Her Majesty's Ministers.* No amount of questions, taunts of evasion and the like would break his resistance to part with more information. Officialdom thus 'washed their hands' on this *loveable pirate,* the greatest frogman in the world—the Briton who blindly fought for and revered with every fibre of his being.

One man believed he was still alive—the author of this revealing book, J. Bernard Hutton, was not convinced that Commander Crabb died beneath the Russian cruiser *Ordzhonikidze*; he was satisfied that the daring frogman had been taken prisoner and transported to the Soviet Union. With the sharpness of a bloodhound, he searched for facts in Britain and behind the Iron Curtains of East Germany, Poland, and into Russia with the help of his very reliable anti-Communist organisation's contacts.

The author's tenacity to root out the facts of Commander Crabb's mysterious disappearance and to throw light onto the workings of the Soviet Secret Service department which employs killers, saboteurs and kidnappers deserves the highest praise. This true story of detection, high adventure, and sudden violence is not a lurid spy-thriller—it is fact.

Preston, 18 December 1967

PROLOGUE

It is doubtful whether Lieutenant-Commander Lionel Kenneth Philip 'Buster' Crabb, O.B.E., G.M., R.N.V.R., would have embarked on his notorious underwater survey of the Soviet cruiser *Ordzhonikidze* had he not met Mr. X; but the 'chance' meeting engineered many months before Crabb donned his frogman suit on that fateful Thursday morning began what later was to become one of the world's most intriguing mysteries.

Ever since the disappearance of the famous frogman there has been much conjecture and speculation about what actually occurred in Portsmouth Harbour on 19 April 1956. 'The Commander Crabb Affair'—to give it its popular title—became an international controversy which has never waned. The British Secret Service and the Admiralty alike have been blamed for the ill-fated spying mission which ended in the 'assumed' death of the frogman. Others have claimed that American agents were behind the venture. In all the arguments and debates in clubs and taverns, in Service messrooms and even in the chamber of the House of Commons, no one seems to have considered the possibility that the Russians might have instigated the hazardous underwater operation to lure Crabb into a trap and kidnap him; yet all the evidence points to the fact that the Russians were the culprits and devised their strategy for a number of reasons.

Lionel Crabb had been a thorn in Moscow's flesh ever since the 'thirties when in China; passing as an embalmer and protected by an influential Chinese bourgeois family he had engaged in active espionage for Chiang

Kai-shek, supplying important intelligence about Soviet-controlled communist actions, political and military manoeuvres and other secret plans. Russian agents in China were instructed to silence him, but before they could carry out their orders he had left the country and returned to England.

The Crabb file lay gathering dust in the Central Index at Moscow Secret Service headquarters for years, but when the wartime frogman became involved with the Polish anti-Soviet underground movement and his activities angered the Kremlin leaders, the Second Directorate* again seriously considered 'eliminating' the dangerous naval officer. Again their intentions were not put into effect. Highest quarters in Moscow decided that Commander Crabb must not be harmed. Potentially he was much too valuable. They had discovered that he was one of the world's most experienced and knowledgeable experts on almost every type of explosive, on revolutionary subaqueous tactics and in all fields of underwater warfare. They decided to recruit this exceptional man; but despite the concentrated efforts of top Soviet State Security agents, they soon realised that they had attempted the almost impossible. Lionel Crabb was not the type of man who could be recruited.

Hell broke loose in Moscow when Russian intelligence agents in Britain reported to Soviet Secret Service headquarters that Crabb had carried out successfully a secret underwater survey of the Soviet cruiser *Sverdlov* while she lay at anchor in British waters in the autumn of 1955. It was decided at a meeting in the Kremlin that immediate steps must be taken to get Crabb into the U.S.S.R. as quickly as practicable. How it was to be done was another matter. The Secret Service directors knew that they would succeed only if the right man,

* The Soviet Secret Service department which trains and employs professional killers and saboteurs and organises kidnappings, assassinations and terrorist acts throughout the world.

with the right cover-story, approached the commander. They began at once the long and tedious task of considering every top agent in Britain to find the man for this particularly tricky job. Their final choice was Mr. X.

Mr. X, a capable intelligence officer and trusted War Office executive, had created admirably the image of a loyal Englishman whose one desire in life was to serve his country but at heart he was a devoted communist, supplying the Soviet Secret Service with classified documents and top-secret information of all kinds. Despite the fact that in his youth he had never concealed his Marxist-Leninist sympathies and though he had been a close friend of Kim Philby, who was known to have been an active communist in his student days, he had not been considered a security risk and had been allowed to climb to a responsible position in an important British ministry.

Through Philby, Mr. X met Donald Duart Maclean and Guy Francis de Moncy Burgess. Their friendship matured and Mr. X eventually became the 'fourth man' in the Burgess-Maclean-Philby set. Finally the subversive quartette decided that there was room for a fifth man, Matthew Smith—ostensibly an American intelligence officer—who at Moscow Secret Service headquarters enjoyed the reputation of being an important double-agent.

The Burgess-Maclean-Philby-Mr. X-Smith quintet continued their cloak-and-dagger life until, in 1949, danger clouds gathered. The British Foreign Office discovered that a secret document had been handed to the Russians. Since there were more than six thousand possible suspects it seemed that the security check might drag on for years. The Burgess-Maclean group were not unduly worried that their connection with the U.S.S.R. might come to light; nevertheless, to eliminate the possibility of focussing suspicion upon themselves, they agreed not to meet in the future but to rely upon sophisticated methods of communication. The new arrangement did not curtail their clandestine activities and, as far as Moscow was concerned, everything went on as before.

The group's almost carefree career as double-agents continued until May, 1951, when Matthew Smith informed Mr. X that Maclean was in serious danger. Smith had discovered that the list of suspects in the Foreign Office affair had been reduced to three—and one of them was Maclean. Kim Philby also brought alarming news. He had found out that Burgess was in considerable trouble and his position endangered; that Maclean was suspected positively of espionage for Russia, that he was under constant surveillance and that his early arrest was being considered.

Confronted with these alarming facts the Soviet control agent enlisted the help of the Soviet Secret Service transport department in getting Burgess and Maclean safely to the U.S.S.R. The task was carried out so efficiently that both men had been smuggled out of Britain before Scotland Yard's Special Branch or M.I.5 could pounce.

The intensified security measures taken in all British government offices following the Burgess-Maclean defection endangered none of the three remaining members of the group. Their wisdom in abandoning meetings with each other paid off. Except for Philby, their links with Burgess and Maclean were unsuspected and they were able to continue their successful work for Moscow. When Philby switched to other espionage duties, Mr. X and Smith were the only remaining members of the original group.

It was Mr. X whom the directors of the Soviet Secret Service chose for the task of playing Commander Crabb into their hands. In the late autumn of 1955, Mr. X's spy leader passed on the Kremlin's instructions, and so it came about that the 'respectable' War Office man began to make discreet inquiries concerning

Crabb's regular haunts and eventually went to a public house in London's West End to meet the commander 'by chance'.

Mr. X found it quite easy to involve Crabb in conversation at the bar. As they drank and chatted idly he was sure that the frogman did not suspect that it was a calculated approach. Crabb, for his part, was quite impressed with the War Office man who was not only good company but, it turned out, an acquaintance of his cousin. The two men met time and again for a drink and at one stage Mr. X believed that he was on the verge of getting a hold over Crabb. But at that point his spy leader stepped in and ordered him to hand over to Matthew Smith.

Convinced that he was on the fringe of success, Mr. X protested; but when he was told that the plan to capture Crabb involved underwater activities, he conceeded that any approach from a War Office man might make the commander suspicious. Army men did not usually concern themselves with naval affairs.

When Mr. X casually introduced Matthew Smith to Crabb, he realised that Moscow had made an astute move. Crabb liked the American from the start and everything indicated that things were moving in the right direction. Crabb and Smith began to meet frequently, though when the commander introduced his new friend to his fiancée, Patricia Rose, she made it clear that she did not like the man and did not trust him.

The announcement that Marshal Bulganin and Nikita Khrushchev would be coming to Britain on a goodwill visit and that they would be travelling in the Soviet cruiser *Ordzhonikidze* was Smith's cue for setting Moscow's trap.

Tapping the newspaper he was holding, he said: 'We should be interested to find out what makes this cruiser the fastest and most manoeuvrable vessel in its class.'

Crabb looked up from his drink. 'What do you mean by "we"?' he asked.

The American laughed. 'Be your age, Crabbie. It's obvious. The C.I.A., the Pentagon, your Admiralty . . . any of the Western brass.'

'I suppose so.' Crabb signalled the barmaid. 'Same again, dear.' He sounded bored.

There was a pause. Then Smith asked quietly: 'Would you be interested if the question of another *Sverdlov* arose?'

'How the hell do I know? It would depend on the Admiralty's attitude . . . '

'You're not thinking straight, old man,' Smith said. 'The Admiralty could never be directly involved, however keen they were to get a detailed underwater survey of the cruiser. Remember, these Russkies are supposed to be coming here on a goodwill trip. Can you imagine the kind of hell that would break loose if it leaked out that the Admiralty was involved in a spying mission on a friendly ship?'

Crabb looked at him steadily. Then he nodded. 'I get the message,' he said.

Smith changed the subject. There was plenty of time yet. The warships had not even left Russia for Britain. Rushing things could only arouse suspicion. He was pleased with the way things had gone so far. He knew that he had kindled the frogman's interest. When he reported to his spy leader he expressed his conviction that the spadework was done. Success might well be in sight.

When it was confirmed that the *Ordzhonikidze* would arrive at Portsmouth on Wednesday 18 April 1956, Smith met Crabb again. This time he told him that he was approaching him officially on behalf of the British Admiralty and the United States authorities. He said the proposed underwater survey of the Soviet cruiser was an Anglo-American project but that officially the Admiralty declined any knowledge or participation in the scheme.

Crabb swallowed the bait and agreed to carry out what he was persuaded had become his patriotic duty. Curiously, despite his years of experience in undercover work he seems to have neglected the elementary precaution of making a few discreet inquiries among his friends at the Admiralty concerning Smith's bona fide. The tale his American friend told must have been supremely convincing.

So the trap was laid by Moscow through the double-agents Mr. X and Matthew Smith. Though nobody ever suspected the truth, there were several pointers to Russian involvement in 'The Commander Crabb Affair'.

As soon as he had completed his mission Smith succeeded in leaving Britain before any security agencies could apprehend and question him. When Mr. X finally came under such suspicion that things became too hot for him, the Soviet Secret Service transport department whisked him out of the country and got him safely to the U.S.S.R. Not long after his escape Smith contracted a serious illness and was slowly dying. Mr. X was rewarded for his service by promotion to resident instructor at the unique Soviet school of Gaczyna, where master spies for the English-speaking world are trained.

1

Journey to Samara

On Tuesday, 17 April 1956, Lieutenant-Commander 'Buster' Crabb was not quite his usual confident self as he walked from his home in Hans Road, Knightsbridge, to Patricia Rose's apartment in nearby Ovington Gardens. His brain was busy with many problems—not least the awkward task of explaining to his financée that in a few hours he must leave London for Portsmouth and that their arrangement to meet some friends during the evening would have to be cancelled.

The day was chilly, with a threat of rain, and he wore a mackintosh and his favourite tweed hat. Walking along, swinging the gold-headed swordstick that could almost be called his trade-mark, he wondered how he could convince Pat that this sudden journey to the coast could not be postponed. He could not tell her that he had agreed to carry out a top-secret underwater survey of the visiting Soviet cruiser, yet equally he could not conjure up any plausible, harmless justification for the trip. In the end, no doubt, he relied on his proverbial luck to carry him through when he faced her. After all, his business partner, Pendock, had asked no awkward questions when presented with the fact that he would be away from London for a day or two. It was equally likely that Pat would accept the situation without to much fuss.

His effort to appear cheerful and nonchalant when he entered Pat's appartment failed. His uneasiness was obvious.

'What's the matter, Crabbie?' she greeted him. 'What's worrying you?'

'Nothing, darling,' he replied. 'I'm fine.'

'There must be something wrong,' she insisted. 'You haven't been yourself for days. It's not like you to be so quiet. Are you ill?'

'It's nothing,' he repeated. 'I've just had a lot on my mind lately.' He lit a cigarette from the stub of the one he had just been smoking and Pat noticed that his usually steady hand was shaking slightly. He said: 'Let's go out of here. I could do with a drink.'

They went to the Grove Tavern, kept by their friend Richard Dipple. To Pat's surprise Crabb ordered a gin and tonic for her and half a pint of bitter for himself. She said: 'What's come over you, Crabbie? I thought you were off beer. Why don't you have your usual gin?'

'I don't feel like a gin,' he said shortly. He carried the drinks to a table and pulled back a chair for her. As he took his own seat he said casually: 'I've got to go to Portsmouth, darling.'

'Oh?' said Pat. It was not unusual for Crabb to be carrying out underwater experiments at the naval base. It did not occur to her he meant to leave the same afternoon.

They drank and talked with Richard Dipple until lunchtime; but when Pat suggested they should have something to eat Crabb replied: 'I don't feel hungry, somehow.'

'You must have lunch, Crabbie,' she said. 'Let's go to the Nag's Head and have some cold beef and a glass of wine.'

They walked to the public house in Kinnerton Street and talked to the landlord, Len Cole, for a while. Crabb ordered the food and a bottle of burgundy, but he ate little and sat moodily chain-smoking. The wine was hardly touched.

'Why are you so worried, darling?' Pat asked again.

'I told you. I'm going to Portsmouth.'

'Yes, I know. But why are you in this horrible

mood?'

'Well . . . I'm going to try out some new gear.'

She laughed. 'You, the ace frogman? Trying out new gear has never worried you before. Why get so worked up about it now? I thought you knew all the answers.'

'Oh, it's just one of those things. But you're right, Pat. It's really an easy job and I ought to be back with you tomorrow evening.'

The girl put down her glass and stared at him. She asked: 'You're going *today*? What about our date with the Barrs? Had you forgotten all that?'

On Thursday evening they were due at a party in the Chelsea home of Commander Barr, R.N. (retd.), and his wife.

'I know, darling. But it can't be helped. I've got to go today.' He looked at his watch. 'Hell! It's past two and I've got to be on the 3.18 from Victoria. We'll have to shift.'

As they left the Nag's Head he said suddenly: 'Come down to Portsmouth with me. I'm dreading travelling alone.'

'Oh, my God! I don't want to go all the way to Portsmouth,' Pat snapped. 'What's the point of it?'

Len Cole had walked with them to the door. He cut in pacifically: 'It's only an hour's journey, Pat. It's not worth making a fuss about.'

'It's all right for you to talk,' she retorted. 'But it's a damn sight longer than an hour—and then I've got to get home again.'

Len shrugged. 'Well, it's your business, but I can't understand why you are getting so worked up. Crabbie's in one of his morbid moods and I think you ought to go with him.'

Eventually she decided to accompany her fiancé.

* * *

The Portsmouth train was half-empty and they had no difficulty in finding a vacant compartment. Crabb huddled in a corner seat, paying scant attention to Pat. He started unseeingly out of the window, smoking one cigarette after another. He seemed entirely preoccupied with the diving job that lay before him.

As the train pulled out of Horsham station he took a wallet from the inside pocket of his jacket, went through the contents and found two photographs. One, a colour print, showed Pat sitting beside a swimming pool in the South of France. The other pictured her with her brother and sister-in-law. Crabb studied the photographs for some time; then, as he replaced them in the wallet and put it back into his pocket, he said quietly: 'I love those pictures. I love you, darling.' He took Pat's hand in a sudden movement 'You *do* love me?'

'Yes, Crabbie. I do.'

He leaned back, smiling. 'We'll get married soon, dearest, and we'll be very happy. We'll have a lovely home in the country. You're going to have absolutely everything. We'll have a wonderful time.'

'Of course,' she agreed. 'But I'm not at all happy about you going to Portsmouth now. What are you up to, Crabbie? Please tell me.'

'I can't,' he said miserably. 'I just cannot. It's something I can't tell a single soul.'

Her composure broke. 'Damn it, Crabbie, I can't stand this mental agony,' she said. 'If you don't tell me, I swear I'll break off our engagement.'

He was startled. 'You can't mean that?'

'I can, and I do. I've taken all I can stand, these past few days.'

He inhaled smoke deeply, still staring at her. At last he said: 'Pat, I'm going to look at the bottom of that Russian cruiser when she ties up at Portsmouth tomorrow.'

'You're going to do *what*?'

'I've got to do it,' he said. 'I've agreed to make the dive.'

'But that's sheer lunacy!' She was almost shouting. 'How in God's name do you think you'll get away with it? The Russians aren't fools. The ship will be guarded day and night. You haven't a chance.'

He avoided her eyes. 'You know that I did it with the *Sverdlov* last year. It was as easy as pie.'

'And don't you think that will make them doubly cautious this time?' she demanded. Then, changing her tone: 'Please don't do it! Don't push your luck. Let's go back, Crabbie. *Please*!'

'I must do it,' he repeated stubbornly.

'But who has asked you to do such a lunatic thing?'

'The Admiralty are sending me.'

'The Admiralty?'

'Well, indirectly . . . '

Pat spread her hands helplessly. 'I don't get it, Crabbie. What do you mean—"indirectly"? When the Admiralty want you to do a job they don't tell you in a roundabout way. They give you direct orders. You know that.'

'Well, it's a very special job. Top secret. The Admiralty mustn't know officially anything about it in case there's a leak or something goes wrong. Very special precautions are imperative, so that no connection between the Admiralty and me can ever be established.'

She shook her head. 'It still doesn't sound right to me. I don't like it at all.'

Crabb tried to reassure her. 'That's because you don't know the ins and outs of how secret work is done,' he said. 'There are occasions when manoeuvres of this sort are vital, and there's no need for you to worry your pretty head unnecessarily.'

But Pat was not convinced. 'If the Admiralty are sending you "indirectly", who is giving you your direct orders?' she wanted to know.

'Matthew Smith. You know him. I introduced him to you.'

'Oh, him,' she sighed. 'I might have guessed it. You

know I don't like him and don't trust him.'

'He's all right.'

'So you keep saying—but the whole thing smells fishy. I don't like it a bit. This man Smith is an American.'

'Well?'

'Since when has the Admiralty been approaching you through some doubtful American adventurer? You've always received your orders from British naval officers—even when the job was hush-hush. Why the sudden change?'

'Darling, you don't know the full story. Smith explained that this was a joint Anglo-American operation. To exclude the possibility that the Admiralty might ever be suspected of having had anything to do with an underwater survey while the Russians were on a goodwill visit in our waters, the approach and all arrangements *had* to be made by the Americans. I think it was a very wise decision. Very ingenious indeed.'

'You may think so. To me it smells fishier than ever.'

'Just because you don't like Smith . . . '

'Who *is* this Matthew Smith anyhow?' Pat demanded. 'How do you know he's telling the truth? How do you know you can trust him?'

Crabb said: 'He is an American intelligence officer.'

'But this is a naval matter. Matthew Smith isn't a U.S. Navy officer, is he?'

'Sometimes naval matters are engineered and carried out by civilian intelligence officers—especially if and when the Americans are involved.'

'Well, I still don't like it. Please, Crabbie, call the whole thing off. There's still time.'

'I can't,' he said desperately. 'I've got to do it. I have given my word.'

For some time there was silence. Pat watched the blue smoke from Crabb's cigarette rising and swirling above his head. She wondered what more she could do or say to make this stubborn man change his mind.

'What are they paying you for this job?' she asked

at last.

'Fifty pounds.'

She looked at him pityingly. 'You must be out of your tiny mind,' she said. 'You are not short of money. You are writing your book for England and America and you know it's going to be filmed. You're making a good living out of your business with Pendock. Why risk your neck for fifty quid? You must be completely mad.'

'I'm not doing it for the money,' he protested. 'It's an important job—and it's a challenge in a way.'

Pat gave up. She realised that nothing would induce Crabb to abandon the dive and his survey of the *Ordzhonikidze*, said to be of revolutionary design and perhaps the most modern warship in the world. Despondent and outraged, she knew there was nothing she could do to save the man she loved from terrible danger.

* * *

The train reached the outskirts of Portsmouth. Crabb stood up and collected his belonging and swordstick from the luggage rack. He told Pat: 'I'll do my very best to be back in London by tomorrow evening and if I don't arrive too late I'll ring you and we'll go out somewhere. If I can't make it, I'll come and collect you for lunch on Thursday.' Then, as they left the compartment at Portsmouth station, he said: 'Thank you, darling, for having come with me.'

'I'm glad, now, that I did,' she said. 'Let's go to the Queen's Hotel for tea.'

'I'm afraid I can't. I'm meeting Smith and you don't like him.'

'I hate him,' she burst out. 'Oh! if only you'd listen to me. If only you'd throw that nasty bit of work overboard.' She was on the verge of tears.

Crabb took her in his arms and kissed her. 'Don't worry, dearest. Thank you again for coming with me.

I'll see you tomorrow evening or on Thursday lunch-time. And once this little job's over and done with, I'll never worry you again, my love. Have a pleasant journey back—and take very good care of yourself.'

Pat's eyes were wet as she watched Crabb's short, stocky figure mingling with the crowd at the ticket barrier. The long train journey back to London was one of the most depressing she had ever experienced.

2

The Trap Is Sprung

Sitting at the table by the window of a café in Portsmouth High Street Matthew Smith drank tea and watched the endless stream of passers-by. Eventually he saw Commander Crabb emerge from the crowd and begin to cross the road. He put some money on the table, left the café, joined Crabb and walked with him along the High Street towards Gosport ferry. Their destination was the Sallyport Hotel, only a stone's throw from the docks area and ideal for their purpose because it afforded a great deal of privacy.

When they arrived at the reception desk and signed the hotel register the proprietor, Mr. Edward Richman, inquired how long they intended to stay. 'A night or two,' Smith said; but Crabb said he intended to leave the following day, if possible.

They were shown to their rooms. As Crabb unpacked his bag the elderly chambermaid, Lily Cripps, knocked and entered the room. 'Good evening, sir,' she said. 'It's a great honour to meet you.'

'You know me?' Crabb asked, surprised.

'Oh, yes, sir. I've heard so much about you that it's almost as if I knew you personally. If you don't mind my saying so, sir, I think your wonderful. I can't tell you how excited I am to see you here at the Sallyport. All my friends will envy me.'

Crabb was amused and flattered by the grey-haired woman's obvious hero-worship. He talked with her about the war and his frogman exploits, of which she had read in the Sunday papers; he cracked jokes to overcome her nervousness, and before long they were laughing and chatting like old friends.

When Matthew Smith walked in, he abruptly ended the gossip which he seemed to consider beneath Crabb's dignity. 'We must be off,' he said curtly. 'There's plenty to be done.'

Miss Cripps flushed. 'I'm sorry to have delayed you, sir,' she apologised. 'Thank you very much for talking to me. I shall always treasure the memory.' As she was about to leave the room she turned and asked: 'Would you like early morning tea, sir?'

'Yes, please,' Crabb said. 'About seven o'clock.'

Smith shut the door behind her and said: 'You had better see to your gear now, so that everything's O.K. when you need it.'

'What about your own gear?' Crabb asked.

'I won't need it.'

'How's that? You said you were coming with me.'

'Plans were changed,' Smith said. 'It has been decided that I am to supervise the operation from the waterfront.'

'H-mmm! That makes the job rather dodgy,' Crabb objected.

'Surely an expert of your calibre doesn't need a supporting frogman,' Smith said acidly. 'Especially for such a trivial operation.'

'Trivial operation, my foot! You don't seem to know what you're talking about. A dive to a depth of thirty-three feet or more with my equipment isn't trivial. It's a hell of a task.'

Smith said: 'If you want to call it off, that's O.K. by me, buddy. If you're scared, we'll pack up and get back to London. But I leave it to you to figure out what people are going to say and think about you.'

'Don't panic. I'll do it.'

'You'd better see to your gear then,' Smith said with unmistakable relief. 'Today—before the Russian ships get here. Tomorrow everything is sure to be more difficult.'

They left the bedroom together. Neither suspected that their conversation had been overheard. Nor—more important still—did Crabb suspect that Matthew Smith was a double-agent working for both U.S. Intelligence and the Soviet Secret Service.

Crabb and Smith parted outside the hotel after arranging to meet later for a meal. Crabb made his way to *H.M.S. Vernon,* the establishment which houses the naval diving school which, through the years, had become almost his second home.

'Hullo, Crabbie,' a naval lieutenant greeted him. 'What's up?'

'Nothing's up. I've just dropped in to say hello to my friends.'

'What a coincidence,' the young man said. 'Just when a Russian cruiser and a few destroyers are on their way to Pompey.' He grinned. 'Don't get near the Russkies when they arrive tomorrow, Crabbie. They're sure to be prepared for trouble.'

'Why the hell should I want to go near them?' Crabb demanded.

'Well, I haven't forgotten the *Sverdlov* episode.'

'Oh! That's ancient history . . .'

The lieutenant's voice expressed friendly anxiety. 'I hope for your sake, Crabbie, that you won't do anything foolish. Getting away with that sort of thing once is an almost incredible bit of luck. To hope to get away with it twice is next to sheer lunacy.'

'Thanks for your concern,' Crabb tried to shrug off his friend's suspicion. 'You can rest assured my arrival in Portsmouth now is what you said—sheer coincidence.'

'I hope so,' the lieutenant repeated. 'I certainly hope so.' He sounded pessimistic.

They continued talking for a little longer and then Crabb went to look up some more of his friends at the school. Eventually he made his way to the divers' room.

The man on duty greeted him warmly and, like the

lieutenant, asked whether he was up to his old tricks. He did not take Crabb's denial with greater confidence than his brother-officer, especially when the frogman began to show interest in the shelves where the oxygen cylinders and diving equipment were stored.

'I'm a bit low on oxygen . . . ,' he began.

His friend cut in hastily: 'Now, Crabbie, you know I'd do *anything* for you, but as things are right now, I can't let you have as much as a whiff of oxygen or anything else unless you produce a special voucher for it. Strict orders, old boy. With those Russian ships arriving tomorrow, everything has to be authorised by the powers that be.'

'I'm getting slightly tired of hearing about those Russians,' Crabb said. His spirits began to sink. If he could not get the oxygen supply for his dive, he would be forced to call off the operation. Nevertheless, he made no attempt to persuade the officer to 'bend' his orders. He knew the man well and was aware that any such move would be useless. He would have to trust to his proverbial luck and hope that he would be able to find the vital cylinders elsewhere.

Then, unexpectedly, his friend said: 'I hope you don't mind if I leave you on your own now. I'm off duty and want to get to the canteen before everything goes off the menu.'

Crabb's heart leapt. Trying to keep his voice steady, he said: 'Of course I don't mind. I might join you for a cuppa later on.'

Alone in the divers' room he took the oxygen he needed, checked his diving gear to make sure that it was in perfect order, and then stowed it in its container. Shortly afterwards, he joined his friend in the canteen and they chatted freely until it was time for the officer to resume his watch.

* * *

Crabb was in high spirits when he left *H.M.S. Ver-*

non. He was now perfectly equipped for his underwater expedition He had studied the official charts, knew the times of the tides and was able to assess the most favourable conditions for entering the water. Everything had turned out very much more easily than he had expected and he felt sure that his luck would continue to hold.

Meanwhile Matthew Smith had also been busy. He had contacted a Soviet shore agent and confirmed that Crabb's planned dive would be made. He promised to keep the man informed about every important move, so that the commander of the Russian flotilla could be advised exactly when to act.

After Crabb had put away his diving gear he met Smith for dinner.

'Everything O.K.?' the American asked as soon as Crabb sat down at the table.

'Everything's fine.'

'That calls for a celebration,' Smith said. 'We must drink to the success of your——' He stopped short as he realised that the waiter was standing near and could overhear every word that was said. He asked for the wine list, and during the remainder of their stay at the restaurant they talked only of trivial matters.

They returned to the Sallyport at an early hour. Crabb had to have a good night's sleep if he were to be completely fit for his exacting job next day. They had their own key to the hotel and nobody saw them let themselves in. Smith's anticipation that the hotel would afford them maximum privacy appeared to be confirmed.

The next morning, Wednesday, 18 April 1956, Lily Cripps brought Crabb his morning tea at seven o'clock sharp. Once more she was delighted by Crabb's cordiality. 'He was a wonderful gentleman, the Commander,' she was to say later.

* * *

By the time Crabb and Smith reached the docks area, the *Ordzhonikidze* and her accompanying destroyers lay alongside South West Jetty. Apart from the fact that more people than usual had come down to look at the warships which had brought Marshal Bulganin and Nikita Khrushchev on their goodwill visit to Britain, things were almost normal.

Crabb and Smith strolled casually along the waterfront like any of the other sightseers, but as they neared the Soviet ships Crabb's eyes were expertly assessing the situation. He decided that conditions were favourable and that his task might not be as difficult as he had anticipated. During his reconnaisance he also decided on the spot where it would be most advantageous for him to enter the water. It was well suited for diving unnoticed, the distance to the cruiser was reasonable, and everything appeared to be in his favour.

When they returned to Gosport ferry Crabb stopped to watch the cars and motorcycles being driven on board. As the ferry left its berth and began to nose across the water he parted from Smith after arranging to meet him later to tell him the proposed time of his dive. He walked on towards *H.M.S. Vernon,* where he hoped to find out as much as he possibly could about the Russian vessels.

His lieutenant friend greeted him as warmly as he had done on the previous evening, but it was plain that he still felt uneasy about the real reason for the frogman's visit to Portsmouth. 'Don't do it, Crabbie,' he warned.

'What are you on about?' Crabbie asked impatiently.

'Don't play about, Crabbie. You know bloody well what I mean.'

'You still think I want to look at the bottom of that Russian cruiser?' Crabb said. 'I told you yesterday that I hadn't come down here for anything of the sort, but you seem to have made up your mind not to believe me.'

'I'm worried about your safety, that's all,' his friend said. 'We know what's up and around these ships. Everything points to it that they expect a repetition of the *Sverdlov* business and they're determined not to let any body get near the vessels. Their look-outs are on round-the-clock duty, their divers and frogmen are alerted and ready to intercept any intruder, and the whole set-up is a deathtrap—even for a man of your unique abilities and skill. That's all I'm telling you. I wouldn't like to see you get into serious trouble.'

At least one other of Crabb's naval friends warned him on similar lines. When he left the underwater establishment he knew that he had not succeeded in convincing anyone that his visit to Portsmouth was 'just coincidental'. Nevertheless, his talks had been more than valuable. He had learned of the precautions the Russians were taking and, forewarned of the dangers, he would be able to take the necessary steps to avoid possible traps and to complete his mission successfully.

* * *

He met Smith again at the café which they had chosen as their rendezvous and told him that he had decided to dive at about seven o'clock the next morning.

The American was furious. 'That's an irresponsible decision,' he snarled. 'It's downright idiocy. How do you expect to get away with diving in broad daylight?'

'It's absolutely safe,' Crabb replied. 'I'm the one who is taking the risk and if I was not perfectly satisfied that it was all right I wouldn't suggest it.'

Smith thumped the table. 'I want to have it over and done with as quickly as possible, and I insist that you do it during the evening or night, when there's the minimum risk. I am responsible for the operation and I am the one who must be satisfied that everything goes smoothly and safely.'

'You can like it or lump it,' Crabb retorted. 'It's

going to be tomorrow morning or not at all. And now I'm off to meet some friends for a drink.'

'So that's it!' Smith said. 'You don't want to dive because you want to get boozed up first. Your nerve's going.'

'Poppycock! I'm having a drink with my friends because there's nothing else I can do at the moment. I'm dependent on the tide, and that's my reason for doing the job tomorrow morning. Now you know the ins and outs, and if you want to join me for a drink you're quite welcome.'

As both men got up to pay the bill, they received a jolt which cooled their tempers. The café proprietor stood nearby, pretending to busy himself, but they realised that he had been listening to their quarrel. They knew that the eavesdropping might have dangerous consequences but they dared do nothing about it. To have said anything would only have made matters worse. They could hope only that the man was just a nosey parker who would quickly forget what he had heard. If it turned out that he had passed information to the authorities and Crabb's dive were stopped, they would have to devise other plans.

Smith went with Crabb to Monck's Bar in the old High Street and met his friends—a petty officer and a lieutenant-commander and his wife. It was a reunion of old companions and the talk was all of earlier days. No one mentioned the Russian warships. The party broke up before midnight and, as on the evening before, Crabb and Smith entered the Sallyport unnoticed.

* * *

While Crabb donned his frogman suit, Smith managed to give the Soviet shore agent the prearranged secret signal that the dive was about to begin. From the agent's acknowledgement he knew that the commander of the *Ordzhonikidze* would receive the awaited high-speed radio message without delay.

As Crabb entered the water and swam submerged towards the Soviet warships, Russian divers went overboard to intercept him. They were under strict orders not to kill but to overpower him and take him through an underwater airlock door in the cruiser's hull.

Unknown to Crabb and Smith and unnoticed by the look-outs on the Russian ships, a British petty officer and lieutenant saw Crabb enter the water and submerge in the direction of the *Ordzhonikidze*. Both officers also saw Russian divers jump into the water and take up strategic positions around the cruiser. Later they watched a frogman surfacing in the vicinity of the cruiser, watched a skirmish between the lone frogman and others and saw them all, after a short, fierce struggle, disappear beneath the water. They kept the *Ordzhonikidze* and the destroyers under continuous surveillance but saw nothing unusual happening on any of the vessels. They were almost certain that all the men had either boarded the flagship through an underwater airlock or climbed aboard one of the vessels at a point invisible from their positions. When they submitted their reports, they added that they hoped the Russians would not execute Commander Crabb as a spy.

The divers from the *Ordzhonikidze*, reinforced by others from the destroyer *Smotriashchin*, succeeded in cutting off Crabb's escape, managed to overpower him and take him through the underwater airlock aboard the cruiser. When they handed over their prisoner to the State Security officers in the flagship, they reported that the British frogman had put up a brave fight but had not had a chance against such an overwhelming force. All the Soviet divers who took part in this operation were rewarded, after their return to the U.S.S.R., for having 'carried out their task to the fullest satisfaction'.

At about the time when the captured Crabb was

brought before the flagship's commander and political officer, a signal was flashed from the *Ordzhonikidze* which informed Smith that his mission had been accomplished. The double-agent understood the message and left his observation post, picking up Crabb's personal effects.

3

Missing . . . Presumed Dead

While Matthew Smith was travelling from Portsmouth to London Patricia Rose was waiting impatiently for Crabb to keep their luncheon appointment. She had remained at home all the previous afternoon and evening, expecting his promised telephone call. She had been disappointed but not unduly worried when he failed to ring her. After all, he had not been too sure whether he would return on Wednesday evening or Thursday morning. But when, at noon on Thursday, he had still not arrived at her home she began to fear that something had really gone wrong. By one o'clock she could bear the oppressive quiet of the apartment no longer. She pulled on a coat and hurried out to the Nag's Head in Kinnerton Street.

When she walked into the tiny saloon bar Len Cole saw that she was fighting hard to hold back her tears. Her face was pale and she was plainly in a bad state of nerves.

'Len,' she said without preliminary, 'Crabbie is missing.'

He poured out a gin and tonic for her. 'Drink this. It will do you good. Now, what's the trouble?'

Her hand was shaking as she took the glass. 'Crabbie's missing,' she repeated almost in a whisper. 'I'm afraid we won't see him ever again. He should have been back in London last night or at the latest this morning.'

'Oh! Come on now, Pat. You know Crabbie,' Cole said soothingly. 'He might walk in here any minute. There's nothing to get upset about.'

She shook her head. 'I know exactly what he was up to in Portsmouth, and it wasn't funny.'

She told her friend about Crabb's plan to survey the hull of the *Ordzhonikidze*. His face grew serious as he listened, but at the end he said : 'Don't worry, Pat. He's too clever to be caught by any Russians. He's probably only delayed the dive. He'll be back as soon as he can.' He tried to put conviction into his voice, but he felt that Pat might well be right and that the frogman's boasted luck might have run out at last.

There were too many people who knew Pat in the bar and she was in no mood for idle chatter. She left the Nag's Head and went to the Grove Tavern, in Beauchamp Place, against hope that Crabb might be there, waiting for her. Richard Dipple remembers that she arrived almost in a state of collapse.

'Richard, give me a drink, please,' she said. 'I'm *so* worried. Crabbie is missing.'

'Missing?' he asked incredulously. 'How do you know?'

'I do, Richard. I know an awful lot. We'll never see Crabbie again.'

'But why, Pat?'

'I can't tell you. I've just got this awful feeling that something terrible has happened. Oh! Richard, I don't know what to do . . .'

He did his best to comfort her but she seemed to be beyond reason. Nothing he could say seemed to reach her.

Her next call was at the Lowndes Arms, in Eaton Place, where she had been many times with Crabb. The first person she saw was her actor friend, Dennis Price. He noticed her distress and asked what was wrong.

'I'm getting frantic with worry,' she replied. 'Crabbie is missing.'

'Surely not . . .'

'I'm afraid he is.'

'Come and have a drink, Pat,' Price said. 'Maybe you are worrying quite unnecessarily.'

'No, Dennis. I have a feeling that I am not going to see him again.'

'He'll turn up, all right,' the actor said encouragingly. 'You'll see. Everything's going to be all right.'

She was grateful for his sincere attempt to put her mind at rest and finally, calmer, she returned to her apartment to be there when Crabb arrived. But as time ticked by and there was still no sign of him, her fears redoubled. She became more and more convinced that she would never see her fiancé again.

Later in the afternoon she telephoned his business partner.

'Look, Pendock,' she began, 'I'm very worried about Crabbie. He——'

'Don't be worried,' he interrupted. 'You know Crabbie. He's probably hiding behind a rock with a pretty mermaid.'

'I'm sorry, but I don't feel like joking. Have you got any idea what he was doing in Portsmouth?'

'No. He said something about testing some new gear but I wouldn't know what he meant. I know nothing about diving.'

'Well,' she said, 'I'm not going to tell you what he was supposed to do, but I don't think we are going to see him again.'

'Don't be so silly! Of course we'll see him. He's probably held up,' Pendock said. 'He 'phoned me last night and told me that he would be away a bit longer than he had anticipated. So, you see, there's not a thing to worry about.'

He sounded so sure that Pat began to hope again that her fear was unfounded and that Crabb would be with her again shortly. But then, she reasoned, he would have been certain to telephone her from Portsmouth if he were all right. And once more the dread returned.

She could stand the solitude of the apartment no longer. She went down to the street, hailed a taxi and was driven to the Chelsea home of the Barrs. The commander and his wife did their best to reassure her and for an hour or two she seemed happier. But when she eventually returned to her own home she sank deeper and deeper into despair.

* * *

While Patricia suffered alone in her apartment Matthew Smith and Mr. X were with their Soviet spy leader. All three were in high spirits now that their important assignment had been completed. They agreed that Crabb's personal belongings should be sent to his mother, but his wallet and its complete contents and his gold-headed swordstick were taken by the spy leader for transmission to his superiors in the U.S.S.R.

By Friday morning, 20 April, Pat was frantic. She had lain awake all night and had watched the first light break over London. No 'phone call. No Crabbie. She was almost out of her mind. She telephoned Crabb's business partner again, repeated her fears, and said: 'Look, Pendock, you know I can't get in touch with the Admiralty, so will you please do something to find out what's happened. *Please* ring them!'

He promised to do so. Before contacting anyone in authority, however, he telephoned the Sallyport Hotel to find out whether the staff there knew anything. He was dismayed when he was told that Crabb had gone and that on the previous day a Mr. Smith had collected his own and the commander's belongings and paid both bills. He immediately contacted the naval authorities but his inquiries produced very little. Accurate information was not yet available, he was told. He would be advised as and when anything was known.

* * *

From the moment he had been captured and taken aboard the *Ordzhonikidze* Commander Crabb had stubbornly refused to answer any question Rear-Admiral V. F. Kotov, Political Officer K. G. Vosensky or Intelligence Officers A. A. Shcherkin and V. B. Dmitriev put to him. Indeed, he had behaved as though he was deaf and dumb. When the officers realised that nothing they could do would make their prisoner talk, he was drugged and transferred to the cruiser's hospital quarters. The commander of the flotilla considered this the most suitable place because, if the British authorities decided to make inquiries, Rear-Admiral Kotov could declare quite plausibly that Crabb was one of his seamen who had fallen ill.

In a masterly piece of bluff Kotov used a conversation with Rear-Admiral Burnett, Chief of Staff of the Portsmouth naval base, to draw his attention to a report from seamen aboard the *Ordzhonikidze*. 'At 7.30 a.m. on 19 April, they observed a frogman floating between the Soviet destroyers,' he told Burnett. 'The frogman, who wore a black diving suit, with flippers on his feet, was seen on the surface of the water for a short time and then dived again alongside the destroyer *Smotriashchin*.'

Burnett categorically rejected the possibility of the appearance of a frogman alongside the Soviet ships and said that at the time indicated there were no operations in the port involving the use of frogmen.

This conversation between the Soviet and British rear-admirals was not made known at the time.

Patricia Rose's anxious remarks to her friends and acquaintances; Pendock's insistence that the naval authorities should provide him with information about what had happened to his business partner; the conversation between Kotov and Burnett, together with other significant incidents, made it plain that sooner or later 'The Commander Crabb Affair' must become a matter of public interest. To prevent newspaper reporters and other interested people ferretting out too many details and possibly piecing the full story together, the authorities decided to take precautions to prevent the mystery from being unravelled.

On Saturday, 21 April 1956, Detective-Superinendent Stanley Jack Lamport, head of Portsmouth C.I.D., went to the Sallyport Hotel and removed four pages from the hotel register, covering the first three weeks in April. He signed the next page as a receipt and warned the proprietor, Mr. Edward Richman, and his staff to say nothing about the movements of Commander Crabb and Matthew Smith. The ban was observed and the removal of the register's pages remained secret for some time.

A week later, on Saturday, 28 April, while the Soviet flotilla steamed towards their home base, Baltiysk—with Crabb lying drugged in the *Ordzhonikidze's* sick bay—a commander from the Admiralty went to see Patricia Rose. She was out when he arrived at her apartment in Ovington Gardens, but he left a message for her with Mr. Lloyd Prides, the Australian landlord of the building.

When Pat returned, Mr. Prides told her: 'A commander from the Admiralty came to tell you that Commander Crabb is missing. He said there is still a remote chance that he might have been picked up by a ship without radio, that he might he safe and that he might turn up some time somewhere.' Both knew that the 'remote chance' phrase was just a kindly attempt by the officer to soften the blow.

On Sunday, 29 April, the Admiralty made the following short announcement about Crabb's fate: 'He is presumed to be dead as a result of trials with certain underwater apparatus. The location was in Stokes Bay, and it is nine days since the accident.'

The official announcement shocked not only Patricia

Rose but also Commander's Crabb's close friends, fellow-officers and many others. There was no longer any pretence that he was 'missing', and although it was disclosed that the accident had occurred well over a week before, so far no attempt appeared to have been made by the naval authorities to find the body.

The most curious sentence in the official statement was: 'The location was in Stokes Bay.' Stokes Bay was some three miles from the spot where the *Ordzhonikidze* had been anchored—an almost impossible distance for Crabb to have swum. When the naval correspondent of *The Times* attempted to obtain the original of the statement, he was told that it was 'not available'.

That same Sunday Commander Blanders, master of a Danish destroyer, observed the *Ordzhonikidze* in the Skagerrak, while she and the two escorting destroyers were on their homeward voyage. The report from the Danish destroyer to base was released officially. It stated:

'Eleven degrees eastern latitude, a helicopter, which did not belong to the two aircraft on board the vessel but which had arrived earlier from an easterly direction, took off. Commander Blanders, master of the destroyer, watched through his binoculars two men who went on board the helicopter. One of these men was being led.'

It was later revealed by the authorities that the radio officer aboard the Danish destroyer had intercepted two coded messages. These were passed to Igor Vutnovko, a former Soviet naval officer and cypher expert who, in 1950, had escaped from the U.S.S.R. and married a Dane. He succeeded in decoding them.

The first, transmitted to the *Ordzhonikidze* half an hour before the helicopter landed, read: 'Landing in thirty minutes at prearranged position to take off prisoner.'

The second, transmitted to a military airfield in the U.S.S.R. a quarter of an hour after take-off read: 'Taken off prisoner. First-Lieutenant of the Navy

Kutzov on board, acting as watch officer. Request fighter protection until target airfield.'

The official Danish announcement added that the requested fighter cover had arrived.

There was little doubt in anybody's mind that the prisoner was Commander Lionel 'Buster' Crabb.

4

Breakfast With Zhabotin

The Admiralty announcement concerning Commander Crabb's presumed death sent scores of newspaper reporters to Portsmouth to interview anybody who be connected even remotely with the mystery. Others besieged Patricia Rose's apartment in the hope of securing her story for their editors. It was common knowledge that she was engaged to the missing frogman and her telephone rang incessantly. British, European and American newspapers and magazines made tempting offers, stepping up their bids to four and even five figures for the exclusive publishing rights.

Pat seriously considered accepting. Big money was involved and she could use it; but the financial aspect aside, she wanted to tell the public all she knew. By doing so, she thought, she might induce the Admiralty to reveal the true facts about what had happened to the commander.

While she was trying to make up her mind with which newspaper to deal, her prospective mother-in-law, Mrs. Beatrice Crabb, telephoned from Oxfordshire. 'I suppose all the newspapers are after you for your story,' she said.

'Yes. It's perfect bedlam.'

'Have you signed a contract with any of them, Pat?'

'Not yet.'

'I'm so glad,' Mrs. Crabb said. 'I couldn't bear the thought that your private life and Lionel's should be dragged through the Press—at least, as long as I am alive.'

'If that is how you feel, mother, I won't give anyone

the story,' Pat promised.

She kept her word. Though persistent newsmen continued to wine and dine her, increasing even further the sums they were willing to pay, they got no more from her than the makings of an uninformative paragraph.

* * *

On Sunday, 29 April 1956—the day on which the Danish destroyer commander had watched two men being taken off the *Ordzhonikidze* and on which the Admiralty had announced that Commander Crabb must be presumed dead—a Russian helicopter flew the heavily drugged frogman under fighter escort to Stettin. From there an aircraft took him to Moscow under escort by six Soviet State Security officers. When the aircraft landed at Vnukovo Airport Crabb was driven at once to Khimky Naval Intelligence Station, where Investigation Judge Colonel I. S. Zhabotin took charge of him.

In the interrogation room the duty officer ordered a guard to bring dungarees for the prisoner, who was still wearing the pyjamas with which he had had been issued in the cruiser's hospital quarters. While Crabb was dressing, a plate of hot *borshch* was brought in.

'Go on, eat, Commander Crabb. We want you to feel comfortable here,' Colonel Zhabotin said in English. 'Don't sit there yawning and staring in front of you.' He ordered the guard to fetch strong black coffee to wake the prisoner up, and then continued: 'Do you admit that you are the British naval commander Lionel Crabb?' When there was no reply, he began: 'Now, look here. We know everything——'

'If you know everything, why ask me?' Crabb retorted.

'Because we want to hear from you what you have to say.'

'Well, I have nothing to say.'

Zhabotin said: 'I would not advise you to take this

attitude. We do not seek to squeeze secrets out of you. All we want is to have a friendly discussion with you so that we can re-check our facts. But if you persist in your foolish attitude I can assure you that we have also ways and means of making you talk.'

A guard brought in the coffee, which Crabb drank without much persuasion.

'That's right. Drink the black coffee, Commander. It's nice and strong and will wake you up,' Zhabotin said. 'That's better. Did you enjoy it?'

'Thank you.'

'Now, what about confirming that you are the British naval officer Lionel Crabb?' Zhabotin persisted 'Go on. Confirm it. You see that we know who you are; otherwise I wouldn't call you by your rank and name. So why not admit it?'

'If you are so sure that I am who you say I am, there's no need for me to say anything.'

'Your way of thinking amuses me, Commander. You haven't once tried to deny that you are you, but you won't formally confirm your identity. I could force you to do so but I have no such intention. Besides, your non-denial actually amounts to an admission.'

At this moment the guard brought in a loaded tray.

'Ah!' said the colonel. 'Here you have a real Russian dish, *beef-Stroganov,* and *kysel* for an after. *Bon apetit,* Commander.'

When Crabb had finished eating, he asked: 'Did you enjoy your meal?'

'Thank you.'

'Now, what about confirming that you are the British naval commander Lionel Crabb?'

There was no reply.

Zhabotin said reasonably: 'Listen, Commander. I went to help you. I don't want to pump you for valuable information; but if I am to help you, you must drop your silly attitude and help me to help you. It may sound strange to you that despite your despicable spying expedition I should want to help you instead of passing you right away to the tribunal for sentence on charges of espionage. But we Russians are not after blood—especially if we know that we had only caught a tool who has merely acted on orders from his superiors. So perhaps you will now see reason and be more communicative than before.

'As I have already said, we have proof that you are the famous British naval commander, Lionel Crabb, also known as Buster Crabb. If you so wish, you can have sight of the *dossier* about yourself, which includes not only extensive data about you, but also some most interesting photographs.'

There was a short silence. Then Crabb said: 'I am Lionel Crabb. But I am a freelance frogman. I have not worked for the Admiralty for quite a time.'

'I am glad that you have listened to my advice and are beginning to cooperate,' Colonel Zhabotin said. 'You say that you have not worked for your Admiralty for quite a time. That may or may not be so. But you attempted to survey the hull of the cruiser *Ordzhonikidze* for the American naval intelligence, did you not?'

'No. I did not.'

'I know otherwise. I also know the true identity of the man who called himself Bernard and Matthew Smith. But let's forget this question for the moment. What, then, is your story of why you went to examine the hull of the cruiser *Ordzhonikidze*?'

'Out of mere personal interest,' Crabb replied. 'Having heard and read a lot about the striking manoeuvring abilities of the cruiser, and being highly interested in naval constructon, I fell to the temptation.'

'You really want me to believe such a story, Commander?'

'I do not ask you to believe anything. I have merely told you the truth.'

'And I tell you that you went to examine the *Ordzhonikidze's* hull on orders from American naval intelligence', Zhabotin said. 'I am making this definitive statement on account of extensive investigations which have been made in this matter. You would be surprised if you knew how much information we have about the whole affair.'

Crabb said: 'All I can say is you must have wrong information. You asked me to give you my side of the story, so I am telling you the true facts.

'I appreciate your British code of honour, Commander. I even understand that as a British naval officer you would prefer to take the blame for your action, rather than involve your superiors. But permit me to tell you that as far as your own Admiralty is concerned, you are already a dead man for them. Your Admiralty today announced that Commander Crabb is presumed dead as a result of certain trials in Stokes Bay.'

'Doesn't this announcement confirm that they did not know anything about me?' Crabb replied.

'On the contrary. It destroys your story that the Admiralty does not know anything about your expedition. If it were true that you went to the *Ordzhonikidze* out of mere private interest and without knowledge of your Admiralty or any other admiralty, how do you explain the presence of the senior naval officer posing as Smith? But apart from this important fact, if your story were true and your Admiralty had no knowledge of your espionage attempts, how then can they today announce that you died as a result of certain underwater trials in Stokes Bay? Listen to what your Admiralty has announced: "Commander Crabb is presumed to be dead as a result of trials with certain underwater apparatus. The location was in Stokes Bay, and it is nine days since the accident." Seeing how things are, it is only in your own interest if you revoke your previous story and tell me the down-to-earth facts, Commander Crabb.'

'All I can tell you is that what I have told you is true,' Crabb insisted. 'How do I know that the Admiralty really made this announcement? Reuter's report from London, which you showed me, could easily have been faked.'

'You must take my word for it.'

'Then you must take my word for what I tell you.'

Still quietly, but more forcefully, Zhabotin said: 'I must warn you, Commander Crabb, that it is not our practice to allow prisoners whom we have caught in the actual act of spying to try to play cat-and-mouse with us. You surely realise the serious position in which you are placed and doubtless know that in our country espionage is punishable by the supreme penalty. Also, in your particular case, no one will lift a finger to try to help you because your Admiralty has already pronounced you dead.'

After a pause he continued:

'You have already admitted that you are the famous British naval commander Lionel Crabb. For the purpose of record I want you now to confirm that it was American naval intelligence which employed you to carry out the examination of the *Ordzhonikidze's* hull, And I must make it clear to you that your confirmation of facts already obtained by us does not worsen your position at all. On the contrary, it greatly helps you.'

'I cannot help you there,' Crabb said. 'All I can do is to confirm that what I did was done out of purely personal interest.'

Zhabotin's expression hardened. For the first time he raised his voice.

'This stubborn renewal of your poor fairytale won't get you far, Commander,' he said. 'We have ways and means of making our prisoners talk. If we so wish, we can inject you with drugs which will make you speak the truth. After having got all the information from you which we need, we can let you rot away in prison or we can shoot you. But we can also reward you—that is, if you cooperate with us. It is entirely up to you to choose your own fate.'

'Thanks for your kind explanation,' Crabb said sarcastically, 'but I have already said what there is to be said and I don't intend to invent imaginery tales.'

'Then there is no use continuing our conversation,' the colonel snapped. 'Before I dismiss you I must tell you, however, that not only have we conclusive proof that you attempted to spy on the *Ordzhonikidze* for American naval intelligence and that your own Admiralty was to share in your discoveries; we have also complete evidence that this was not your first attempt to spy on our vessels while in British waters. We have conclusive evidence that you did this kind of thing before, that you carried out underwater spying examinations in our cruiser *Sverdlov* when it lay at anchor at Portsmouth while on a friendly visit to England, and that in fact you managed to get away with it. Now, knowing that we are fully aware of everything you did, you may perhaps find sufficient time to think about whether you wish to cooperate with me or not, while in prison.'

That was the end of Crabb's first interrogation. He was taken to Moscow's *Lefortovo Prison* to endure solitary confinement, remorseless interrogations, brainwashing, punishment and more interrogations.

* * *

Meanwhile, in Britain, newspapermen were beginning to piece together a considerable part of the frogman's story. They told their readers about Crabb's movements up to the time he had left the Sallyport, theorised about the true identity and role of the myterious Matthew Smith, and reported that Detective-Superintendent Lamport had removed four pages from the hotel register. They concluded, too, that the Admiralty's announcement about the experiments in Stokes Bay was a smoke-screen. It seemed much more likely that Crabb had been orded to investigate certain equipment built into the hull of the 12,000-ton *Ordzhonikidze* and had died in the attempt.

On Friday, 4 May 1956, the Soviet Embassy in London delivered a diplomatic note to the British Foreign Office, which stated:

'During the stay of Soviet warships in Portsmouth, at 7.30 a.m. on April 19 seamen on board the Soviet ship observed a frogman floating between the Soviet destroyers. The frogman, who wore a black diving suit, with flippers on his feet, was seen on the surface of the water for one to three minutes and then dived again alongside the destroyer *Smotriashchin*.

'The commanding officer of the Soviet ships, Rear-Admiral Kotov, in a conversation with the Chief of Staff of the Portsmouth naval base, Rear-Admiral Burnett, drew his attention to the case. Rear-Admiral Burnett categorically rejected the possibility of the appearance of a frogman alongside the Soviet ships and stated that at the time indicated there were no operations in the port involving the use of frogmen. In actual fact, however, as it transpired from reports in the British Press on April 30, the fact that the British naval authorities had carried out secret underwater tests in the area where the Soviet ships were anchored at Portsmouth was confirmed. Moreover, these tests resulted in the death of the British frogman.

'It is sufficient to recall that the newspaper *Daily Sketch*, in a brief article on the end of frogman Crabb, reported the following: "He dived for the last time in Stokes Bay, at the spot of secret diving operations near the anchorage of the Soviet ship *Ordzhonikidze*."

'Attaching great importance to such an unusual fact as the carrying out of secret frogmen tests alongside Soviet ships on a friendly visit in the British naval base of Portsmouth, the Embassy would be grateful to the British Foreign Office for an explanation on this matter.'

The British authorities decided to keep the Soviet

move secret from the Press for the time being; but there was a 'leak' and some newspapermen tried their luck at the Soviet Embassy. Mr. Gordon Holman, the *Evening Standard's* naval correspondent, was particularly fortunate, for a Soviet assistant naval attaché told him: 'A watchman in our ship saw a frogman come to the surface. He was on the surface only for a few seconds, then he disappeared and was not seen again. The watchman reported to the captain of the warship.'

Mr. Holman asked whether any action had been taken against the frogman and was told: 'Absolutely none. We were in a British port and there was nothing we could do.' The spokesman said that he did not know on which day the frogman had been seen; that the captain did not report the incident to London; and that the information had just arrived from Moscow.

Captain John Grant, commanding officer, *H.M.S. Vernon*, said: 'If a frogman were seen near the Russian cruiser he certainly did not come from here.'

Among the most popular rumours and theories in circulation were:

1. Commander Crabb was killed by the Russians, who mistook him for a saboteur operating beneath the hull of the *Ordzhonikidze* in Portsmouth Harbour. His body was found floating in the water and was secretly buried at midnight in a local cemetery with full naval honours.

2. Crabb is alive. He was captured by the Russians, kept prisoner in the cruiser and taken back to Russia.

3. Crabb was forced to surface near the Soviet ships by failure of his oxygen breathing apparatus, captured and taken back to Russia as a suspected spy.

4. Crabb was carrying a geiger counter, the size of a fountain pen, which could detect any atom bomb hidden in the underwater magazines of the Soviet ships. He was detected and killed.

5. Crabb was trapped by the cable of a portable television camera while photographing the hull of the

Russian cruiser and drowned.

6. It was a sensational publicity stunt which went wrong.

Because these rumours and theories were circulating not only in Portsmouth but throughout Britain, the Admiralty and political authorities discussed the wisdom of issuing an official statement. They decided against it and the Admiralty consequently added nothing to their previous announcement; but neither did they confirm or deny anything.

* * *

Mr. Sydney James Knowles, B.E.M., who had served with Commander Crabb during the second world war and who'd been Crabbie's closest friend, was determined to find out what had happened. He prepared to dive into the murky waters of Portsmouth Harbour to search for his missing friend, saying: 'I shall never rest happy until I find Buster Crabb's grave.' But a short time before he planned to enter the water, a serving naval officer approached him and said:

'Don't dive. Crabbie is not there.' He added that he knew exactly what had happened to Commander Crabb on the day after the *Ordzhonikidze* anchored at Portsmouth, and went on: 'I know an awful lot about this, but I am sworn under the Official Secrets Act not to say anything. I heard about your attempt to find Buster. It is a dangerous job and I came here to stop you because I should hate to see any friend of mine or Buster sticking his neck out when I know there's nothing to gain. I know Crabbie is not down there. Don't ask me why, but I do.'

So Knowles abandoned his dive.

'My officer friend and every diver I have met at *H.M.S. Vernon* are convinced that Crabbie is alive,' he said later. 'I have a feeling that he's still alive and is being hidden by the organisation that sent him to dive. It's just supposition, but . . .'

5

Trouble in the House

A storm broke loose when Prime Minister Sir Anthony Eden made a short statement in Parliament on Wednesday, 9 May 1956, about Commander Crabb's death. *The Times* carried the following report:

'A fruitless Opposition attempt was made yesterday to get the Speaker's permission to move the adjournment of the House, after the Prime Minister had steadily refused to enlarge on his short statement about the death of Commander Lionel Crabb, the frogman who was reported missing after a dive in Stokes Bay, near Portsmouth Harbour, on April 19.

'Although Labour members received with sounds of shocked surprise Sir Anthony Eden's decision, in the public interest, not to disclose the circumstances in which Commander Crabb was presumed to have met his death, the House heard in almost complete silence the news that what had been done was without the authority or knowledge of the Ministers and that appropriate disciplinary steps were being taken.

'The Prime Minister's firm refusal to be drawn by taunts of evasion and the like was firmly applauded from the benches behind him, but the Opposition showed themselves increasingly restive at their resistance to their pressure for more information.

'The subject arose on Mr. J. Dugdale's question asking for the evidence on which the Admiralty officially presumed the death of Commander Crabb; what were the circumstances of his disappearance; and whether efforts were still being made to locate the body.

'Sir Anthony Eden replied: "It would not be in the

public interest (Opposition cries of Oh!) to disclose the circumstances in which Commander Crabb is presumed to have met his death. While it is the practice for Ministers to accept responsibility, I think it is necessary, in the special circumstances of this case, to make it clear that what was done was done without the authority or the knowledge of her Majesty's Ministers. Appropriate disciplinary steps are being taken."

'Mr. Dugdale said this was one of the most extraordinary statements ever made by a Prime Minister. It was a complete evasion of ministerial responsibility. Mr. Dugdale then put these questions: Why was Commander Crabb diving in the close vicinity of the Soviet cruiser here on a friendly visit? Under whose authority was a police officer sent to the hotel at which Commander Crabb was staying with another man, ordering the leaves of the register showing their names to be torn out? What was the name of the other man? Why did the police officer threaten the hotel proprietor with the Official Secrets Act?

'The Prime Minister replied: "I thought it right to make the statement which I have made, and I have nothing to add to it." (Ministerial cheers.)

'Mr. Gaitskell, leader of the Opposition, said: "The Prime Minister will be aware that a great deal of information has already been published in the Press. Does he not think on reflection, in view of the amount of speculation which undoubtedly will continue in the absence of any information from the Government, that it would really be wiser in the general interest, if a fuller explanation were given?" (Opposition cheers.)

'Sir Anthony Eden replied that he had given the fullest consideration to the matter. "I can assure Mr. Gaitskell," he went on, "that there are certain issues which are the responsibility of the Prime Minister himself, and having given all reflection to all the information at our disposal I thought it my duty to give the answer I have, and I am afraid I must tell the House that I cannot vary from the answer."

'Describing the answer as "totally unsatisfactory to the Opposition," Mr. Gaitskell asked: "Is the Prime Minister aware that while we would all wish to protect public security, suspicion must inevitably arise that his refusal to make a statement is not so much in the interest of public security as to hide the very grave blunder which had occurred?"

'The Prime Minister rejoined that the House and the country "must draw their conclusions from what I have said" (Opposition cries of "It will!") "and also from what I declined to say. Mr. Gaitskell," he went on, "will understand I have weighed up these considerations and they have weighed heavily in my answer. With his experience he knows that there are some of these decisions only a Prime Minister can take, and I am convinced after the most careful reflection the decision was the right and only one."

'Mr. Gaitskell then asked: "Are we to take it, in the absence of any further statement from the Prime Minister, and in the light of what he has just said about the public drawing their own conclusions, that in fact officers or an officer of her Majesty's Forces was engaged upon the business of espionage during the Russian visit?"

'Sir Anthony Eden: "Mr. Gaitskell is perfectly entitled to put any wording he likes on what I said. My words stand as they are without any gloss anyone can put on them."

'Replying to Mr. Shinwell, who asked against whom the Prime Minister was taking disciplinary action and for what reason, Sir Anthony Eden said: "What I said was that disciplinary steps are being taken."

'Mr. Shinwell asked if action was to be taken against some individual or individuals who gave instructions to Commander Crabb; and if it was because they defied authority, or because they acted without consulting Ministers. The Prime Minister repeated: "I have nothing to add to my answer."

'Mr. Dugdale sought leave to move the adjournment of the House, in view of the Government's failure to give a satisfactory explanation of the events connected with the disappearance of Commander Crabb.

'The Speaker, giving his ruling, said: "This application is governed by authority. When a Minister refuses to answer a question on the grounds of public interest it has been ruled in the past—and I adhere to it—that it is a matter which cannot be raised, and therefore I must decline the application...." '

* * *

When the Admiralty was asked whether the information concerning Commander Crabb's death was still as announced on 29 April 1956, the Chief of Naval Information, Captain A. W. Clarke, said: 'The Admiralty has no comment, no further information, and is not prepared to answer any questions at all.'

On the same day Britain replied to the Soviet note of 4 May 1956. The text of the British note was as follows:

'As has already been publicly reported, Commander Crabb carried out frogman tests, and, as is assumed, lost his life during these tests. The frogman who, as reported in the Soviet note, was discovered from the Soviet ships swimming between the Soviet destroyers, was in all appearances Commander Crabb. His presence in the vicinity of the destroyers occurred without any permission whatever, and her Majesty's Government express their regret for the incident.'

* * *

On Monday, 14 May 1956, 'The Commander Crabb Affair' was raised once more in the House of Commons. The Parliamentary correspondent of *The Times* reported:

'The Prime Minister tonight ended his short speech on the case of Commander Crabb, the frogman, by deploring the debate raised by the Opposition and he was rewarded by a long, loud cheer from the benches behind him. From the Opposition came a low growl of dissatisfaction, which carried some cries of "Resign!" . . .

'To mark their disapproval of the "misconceived and inept operation," and in protest against the Prime Minister's refusal to answer any of their questions, Mr. Gaitskell tonight moved the reduction of the token vote on which the debate was held. The motion was negatived by 316 votes to 229.

'The Prime Minister refused to amplify his statement of last Wednesday though he explained the reasons for what he had not said then. He based his attitude on the national interest of security, but in a sudden quiet passage at the end he invoked also the international interest.

'With great vehemence he said that all he cared for was that the outcome of the discussions with the Soviet leaders should, in truth, prove to be the beginning of the beginning. "I intend to safeguard that possibility at all costs," he declared. "I believe it is also in the minds of the Soviet leaders. It is for that reason that I deplore this debate and will say no more."

'His earlier words in a crowded, expectant House with the First Lord of the Admiralty and the First Sea Lord up in the distinguished strangers gallery, and Sir Winston Churchill an attentive listener below, were in firm defence of his reticence. He told the House bluntly that they were dealing with circumstances in which no government in any country would say more than he was prepared to say.

'This warning reduced the House to silence. The Prime Minister explained that there was nothing contrary to parliamentary practice in his attitude and there was no dispute about the general principle that there were certain things which it was against the national interest to disclose. He recalled that Mr. Gaitskell had spoken very freely about the Secret Service and had speculated about their control, organisation and efficiency.

'Sir Anthony Eden said he was not prepared—and he apologised for having to say it—to discuss these matters in the House. He reflected that it was easy for Mr. Gaitskell to imply that all was not well. "I could not answer," he said, "because I could not answer him generally or in detail without disclosing matters which, as he recognises, must remain secret."

'It had never been the practice to discuss these matters openly in the House. "I am not prepared to break that precedent," said the Prime Minister. He insisted that it must be left to the discretion of Ministers to decide these matters. A Minister could not disclose the reasons for his decisions. If he did he would be disclosing what he judged was contrary to the public interest. That was the position in this matter.

'With great deliberation, a pause between each word, the Prime Minister said: "Therefore on this aspect of the matter I must tell the House now I have not one word more to say than I announced on Wednesday." The Government side raised a subdued cheer and the Opposition murmering bore some cries of "Shame!"

'Sir Anthony Eden explained the exceptional course he took on Wednesday of making it plain that what was done was without the authority of Ministers, and he made it clear that that included all Ministers and all aspects of the affair.

'He felt it necessary also to let it be known that disciplinary steps were being taken. It showed that the Government was determined that the proper measure of control and authority should be exercised by Ministers in all matters of this kind.

'The only other matter which the Prime Minister dealt with was the fact that he did not inform the House on Wednesday of the Government's receipt of the Russian note and their reply to it. He denied that he had been holding back information which the House should have had. The Soviet note was delivered to the

Foreign Office on Friday night and in the Foreign Secretary's absence he approved the answer to it on Wednesday morning.

'He claimed that there was nothing inconsistent in that reply and his statement to the House on Wednesday. He realised that the Soviet Government might publish both the note and the reply, but even so it would not have been possible for him to communicate either the effect or text of the note in advance of the reply being received.

'Mr. Gaitskell had combined in his opening speech a measure of restraint and care for security with an effect of probing for the truth. He based it on the constitutional principle that a Minister should take responsibility for what had happened and he deduced that other departments, apart from the Secret Service, appeared to have been involved. But nobody, he said, could claim that these operations were successfully secret and there had been some embarrassment to foreign relations.

'He fully accepted the Prime Minister's disclaimer of ministerial knowledge or approval and said that it should be accepted by the Soviet Government as well as complete evidence of good faith. He did not believe that this episode would do permanent damage to our relations with the Soviet Government, who were realists in this matter.

'He was more concerned with what appeared to be the situation in the Secret Service, because what had been suggested by the Prime Minister's original statement revealed a very grave lack of control at home and a most unsatisfactory state of affairs within the service.

'Presumably the Secret Service and the Admiralty must have been mixed up in the plan from the start. An extraordinary feature of the whole business was the question of how Commander Crabb approached the Soviet vessels if it was the Admiralty's responsibility to guard them.

'Referring to the reported episode of the removal of

pages from a hotel register, Mr. Gaitskell said it was hard to see by what right the police officers made the hotel keeper break the law.

'Mr. Gaitskell concluded that it was impossible for the nation to pass final judgment, but an impression had been created of a most deplorable lack of co-ordination and control between the Foreign Office, the Secret Service and the Admiralty, and an impression of unusual technical incompetence.

'Among the back-bench speakers a Conservative, Lieutenant-Colonel Cordeaux, said it would be wrong for the House to lose faith in the Secret Service because of this case, but it was impossible to excuse it. It was approved as mistakenly and rashly as it was ineptly carried out and he felt alarmed for the higher direction of whatever service might be concerned. Commander Crabb was of an age when he should hardly have been chosen for such a hazardous operation. But Lieutenant-Colonel Cordeaux was convinced that the Russians would attach very little importance to this episode.'

The debate by no means put an end to the efforts which newspapermen and others made to try to solve the deepening mystery.

* * *

On Tuesday, 15 May, Patricia Rose left London for the South of France. Realising that she was dangerously near a nervous breakdown, her brother had telephoned her a few days earlier from his villa in Biot, insisting she should join him and his wife immediately. There was nothing that she could do in London, and she knew that a change of environment could only benefit her; yet, as the train sped towards Dover she wondered whether she was doing the right thing. She played with the thought of returning, to be on the spot in case something important occurred, but soon discarded the notion. The South of France was not the end of the world and, if necessary, she could always fly back to London.

6

Point Of No Return

In Moscow Investigation Judge Colonel Aleksei Feofilovich Myaskov had taken over from his colleague, Colonel Zhabotin, and was in charge of the Crabb case. On Tuesday, 22 May 1956, he had the frogman brought from *Lefortovo Prison* to Khimky Naval Intelligence Station. As soon as the prisoner faced him in the interrogation room, he asked: 'What have you decided to do, Crabb? Are you prepared to answer my questions or are you still stubborn?'

'What do you want to know?'

'Tell me what you discovered when you examined the hull of the *Ordzhonikidze*.'

'Nothing. As soon as I approached the cruiser and applied my measuring instrument, I heard the alarm device. Realising that the game was up, I did my best to get away.'

'So you found out that we don't take chances, eh?'

'I did.'

'Why didn't you throw your incriminating measuring instrument away while swimming to safety?'

'What good would it have done?' Crabb asked. 'If I was caught it was irrelevant whether I had the instrument or not.'

'You have vast experience as a naval frogman?'

'I have.'

'And you can carry out almost any underwater job?'

'I think so.'

'Suppose you had not been caught and were still in Portsmouth, and the Navy asked you to undertake another underwater examination of a Soviet vessel, would you refuse to do it?'

'I don't think so,' Crabb said candidly. 'But I suppose next time I would be more careful.'

'In other words, you do not regret what you did?'

'I regret that I was caught.'

'You realise, of course, that what you say amounts to an admission of spying?'

'I do. But it is not my code of honesty to deny what I did.'

Myaskov stared at him thoughtfully. 'You are a strange man, Crabb,' he said. 'You don't even try to dish up a story that all you did was an innocent morning swim.'

'I don't believe in lies.'

'You know, of course, that we can shoot you for spying?'

'I do.'

'Would you consider buying your life by giving us information about Allied naval secrets?'

'I would not. I am not a traitor to my country.'

'Yet your country betrayed you,' Colonel Myaskov said. 'Your country declared you dead without having made the slightest move to ascertain whether this were so or not. But let's not elaborate on that theme. You don't want to agree to give us information on Allied naval secrets—but don't be surprised when I tell you that all you know we know, too. We know probably much more than you do. We have excellent men everywhere.'

'What do you need *me* for, then?' Crabb retorted.

Myaskov lost his temper and hit the table with his fist. 'You know that some almost perfect reports need clarification in one way or another,' he shouted. 'And you are the man who could help us with it. This is why we did not drown you on the spot but brought you to Moscow instead.'

'I am afraid I cannot help you.'

'You damned silly English fool!' Myaskov exploded.

'Don't you realise that once we have squeezed all you know out of you we can do away with you?'

'Why haven't you done it already?'

'Because we did not want to destroy your brain. We may have further use for you. But if you continue your damned silly nonsense we will do what I said. I ask you now for the last time: Will you do as I ask you or not?'

'I will not,' Crabb said. 'Drug me if you want to. Shoot me. But you won't persuade me to betray my country.'

Again Myaskov looked at him long and hard. After a while he said: 'You know, Crabb, I admire your British spirit. Have a papiros.' He offered his prisoner a Russian cigarette and lit it for him.

'Thank you,' Crabb said.

'You have at least proved that you are loyal to your country, if nothing else,' Myaskov said. 'This, of course, does not erase the crime you have committed against *our* country, where this is the most serious crime which can be committed.' He paused. 'I must tell you, however, that we do not like to rub out people who are valuable in one way or another—that is, if we can help it. And you, with your vast experience as a naval frogman, are doubtless of certain value.'

He paused again, then went on deliberately: 'I am therefore giving you the chance to join the Red Navy as a frogman instead of passing you on to the Military Tribunal for sentence. You would be paid according to scale, which, in your case, would be the equivalent of about £1,000 a year. What have you to say to that?'

Crabb was too surprised to reply. He sat for a while, thinking over the unexpected proposition. At last he asked: 'Provided I agreed, would that mean you would order me to make underwater surveys of British or other Allied vessels?'

Myaskov made a gesture of annoyance. 'Look here, Crabb,' he said. 'I am giving you a chance to save your life and earn your living honestly by following your profession—and you start to make conditions! You are not in a position to make conditions. You either accept what I am offering you or take the consequences. So what is your answer?'

'I can answer only if I know what my duties would be,' Crabb insisted.

'Your duties will be what they will be,' Myaskov shouted. 'How can I know what your commanding officer would order you to do?' Then, more quietly: 'We can make use of you, but we can also do without you. And I tell you here and now: If you decide to remain uncooperative we will not feed you to keep you as a museum piece. We will rub you out.'

Once again he paused. Then he said, 'You know, Crabb, as far as your own authorities are concerned, you are dead. In order to prove their theory we are quite willing to let them find your body—or, to make it quite clear to you, a body which would be accepted as yours. When that was found, the dead naval commander Lionel Crabb could be buried in England while the live Crabb could work undetected and unsuspected in our navy.'

He got up from his chair, crossed the room and opened the door. 'All right. Bring it in,' he ordered.

Two guards entered, carrying a stretcher covered by a rubber sheet. They lowered it to the floor.

'Now, let's see what we have here,' Myaskov said. He turned back the sheet, revealing the body of a dead man. 'This one should fit your description. Don't you agree, Crabb?'

'I'm afraid I don't. I can't see the slightest resemblance to me.'

'The body measurements are nearly enough the same as yours and that's all that matters. As an experienced frogman you ought to know that if a body is left in the sea long enough, it decomposes to such an extent that not even the greatest experts could recognise it or even establish the cause of death. But in order to enable your people to establish your identity the body would naturally have everything you wore when you went out on your spying attempt against the *Ordzhonikidze's* hull.'

He ordered the guards to take the stretcher and body away.

When they were again alone in the room, he continued: 'You see now that I not only mean to give you the chance to join our Navy. I am even prepared to do everything in my power to help your authorities to establish your death without doubt. But don't think you can continue to play cat-and-mouse with me any longer. If in your stupidity you still refuse my offer, you won't leave this station alive. And if that should be the case, your real body will be washed up somewhere near Portsmouth.' He looked searchingly at Crabb and asked: 'What is your answer now?'

The silence dragged into minutes. Then Crabb said slowly: 'Under the circumstances, I accept.'

'Good! I knew that even you, Crabb, had your price,' the colonel said contentedly. 'Your decision does not come as a surprise to me. But I must warn you to give up any hope of double-crossing us. You can take it from me that there is no escape from us. On the other hand, if you work honestly and to the best of your abilities, you will be treated like any other member of the Soviet Navy. We appreciate good work and initiative, and those who distinguish themselves are honoured to the full.'

The confrontation ended. Crabb was driven in a prison van to the rehabilitation centre of the State Security Committee at Selskoye and taken immediately to the camp commander, Klim Osipovich Krasenko.

'I know all about you,' the commandant said in halting English. 'There is therefore no need for you to tell me anything. You have been brought here to acclimatise yourself to Soviet life. The first thing you will have to do is to learn Russian. You must work hard so that you do not waste precious time. I will study your daily progress reports. Your life has now been completely changed for you. Therefore you will forget who you are. For the next ten years you have enlisted in the Red Navy and from now on you will be known as Lev Lvovich Korablov. Your salary will be according to scale from the day you take up your duties. And if you prove yourself reliable you may be allowed to go wherever you wish after your ten year's service is at an end.

'I must add that all that has been said naturally applies only if your record is clean in every respect. I want you to be clear about this. As long as you are in the Red Navy, either in the Soviet Union or in other territorial waters, there will never be a chance for you to double-cross us or to try to escape from us. You will be watched day and night—and let it be said that at the first wrong step you take, you will be shot. There will be no second chance.'

He paused to let his words sink in. Then he continued:

'Investigation Judge Colonel Myaskov is fully aware of the fact that he is not only taking a very great risk but also tremendous responsibility in giving you the opportunity of joining the Red Navy instead of having you shot as a despicable spy. With many saboteurs and wreckers Colonel Myaskov has the reputation of being one of the most ruthless investigation judges; but you, Commander Crabb—or, to call you by your new name, Lev Lvovich Korablov—have discovered by your own experience that he is a very fair man and always ready to give a prisoner a chance. But don't think he is a man who would ever forgive anyone who made the slightest attempt to trespass on his generosity. Anyone who takes the slightest step in the wrong direction finds out that he can also be completely merciless. Always remember my well-meant words of today. If you behave as is expected of any other law-abiding citizen and if you work to the best of your abilities, then, Lev Lvovich Korablov, you will soon find that you have made the right decision and that you have chosen the finest career in the world.'

With this Crabb was dismissed and handed over to Sergeant Vira Nikolayevna Savelyeva, who took him to Barrack VII, where he was to share a room with three other inmates of the State Security rehabilitation centre. The room was furnished with four bunks, a bench and a table.

Before leaving him, the pleasant-looking Savelyeva told him: 'I am now your Russian teacher. You must listen to what I tell you. You must work hard. Here in the Soviet Union we work to plan. Your plan is to learn Russian very quickly. The tovarich commandant will be studying my reports about you. You must work hard or he will be a very furious major. I don't want that. He could be furious with me also. So you must learn very well.'

The newly created Lev Lvovich Korablov assured her that he was very eager to learn Russian as speedily as possible, especially under such a charming teacher. Ignoring the compliment, she told him that his first lesson would begin at 9.30 next morning and would last until 6.30 p.m., with an hour's break for lunch.

As Crabb stretched out on his bunk, he wondered what sort of life lay before him.

* * *

Myaskov lost no time in having the body of the dead Russian prepared for a successful hoax on Britain. His decision to plant a 'man who never was' was made not for the reason he had given Crabb but to deceive the British authorities and, if this grisly plan succeeded, to prove to himself that he was superior to Western intelligence officers.

The body was taken to the Special Commission of the State Security Committee in Leningrad to be made ready for the operation, and on Monday, 28 May 1956, State Security Major Boris Markovich Smirnov sent the investigation judge the following message:

'Body is now ready for immersion in the sea. On advice from medical experts, head and both hands have been removed to exclude all possible identification. Medical experts say that the body must be submerged in the sea for at least one and a half years to decompose to such an extent that no doctor in the world can establish either the cause of death or any possible identification marks.

'The body, bearing all the clothes which the prisoner wore when captured, including this measuring instrument, and the prisoner's frogman suit, inclusive of flippers, has been submerged in the sea since 15.10 hours. To obviate losing the body, indestructible steel wire has been carefully fixed to its legs, so that it is now safely anchored in the sea. To exclude unauthorised people accidentally discovering the body, I have ordered the area to be fenced in.'

Colonel Myaskov was pleased with this report. He felt that everything was going according to plan; but when he received East German and other Soviet satellite newspapers carrying reports on 'The Commander Crabb Affair,' his mood changed. It was not the stories in the Iron Curtain papers which annoyed him. He was quite content that the editors should follow the lead of *Pravda* and *Izvyestia* and speak of 'a shameful operation of underwater espionage directed against those who came to England on a friendly visit'. But the newspapers had also carried two photographs of Patricia Rose, captioned: "The beautiful blonde fiancée of Commander Crabb." One showed her sitting beside a swimming pool. The other pictured her between a man and a woman in riding clothes. Where and how had the reporters obtained those pictures?

It did not take the investigation judge long to find out from Matthew Smith's spy leader in London that the photographs could not have been obtained from any commercial agency. They were plainly amateur snapshots. The spy leader's opinion was that Smith had taken them from Crabb's wallet and sold them to the newspapers. He was emphatic that the pictures had not been in the wallet when Smith handed it to him and he confirmed that he had immediately transmitted the wallet, untouched, to Moscow.

Myaskov at once took steps to reprimand Smith and to call in both prints and blocks from the newspapers. It was he who was directing every move in the Crabb case and he was determined that nobody else was going to interfere.

7

A Message For Patricia

Patricia Rose had always been very fond of the South of France. She loved riding, sailing and swimming and enjoyed parties and visits to casinos. Living with her brother and sister-in-law at Biot she soon regained her former health and peace of mind.

A fortnight after her arrival a family friend, Tony Kessler, an Austrian cavalry officer who lived on the Riviera, telephoned her.

'You wouldn't believe it, Pat,' he said, 'but I'm looking at pictures of you in an East German paper. There's one of you by the swimming pool and another of you standing, in your flowered dress, between Naomi and James in riding kit.'

'You must be joking,' Pat exclaimed.

'I am not. Bring Naomi and James over to lunch tomorrow and see for yourself.'

They went to Kessler's villa next day, and when they looked at the newspaper illustration of the group, Naomi said :

'That's the picture we had taken at Cap Ferrat. Do you remember that Sunday, Pat? We were going off to the show jumping and you were wearing the flowered dress you liked so much. The other snap was taken at our swimming pool.'

Pat looked long at the pictures. She had last seen them when Crabb had taken them from his wallet on the train journey to Portsmouth. She knew there had been only three prints of the group. Naomi had one; Mademoiselle Somborne, who had taken the photograph, had another; and the third she had given to

Crabb because he had asked for it. There were only two copies of the photograph taken by the pool—a colour picture—and one of those had been the one she had given to Crabb.

'How *could* the Press have got hold of those?' she asked. 'Have you still your prints, Naomi?'

'Of course. They're in the album.'

'Perhaps Mademoiselle Somborne gave somebody the pictures,' Tony Kessler suggested.

'She couldn't have done,' Pat said. 'She never had the colour shot of me at the swimming pool. James took it. All I know is that Crabbie had both pictures in his wallet when we went to Portsmouth.'

'Maybe the Russians got the pictures from him,' James said. 'Perhaps it's true that he's alive and a prisoner in Moscow.'

'But he'd hardly have had the pictures with him when he went into the water,' Kessler objected. 'You don't carry a wallet in a frogman's suit.'

'True,' James agreed. 'But both the wallet and Crabbie's swordstick have gone missing. They were not returned to his mother with the rest of his stuff.'

'Crabbie had both those pictures with him when he went to Portsmouth,' Pat repeated. 'I think that ghastly man Matthew Smith got hold of the wallet somehow and gave the pictures to the newspapers. I never trusted him.'

'Well, whatever happened, I can't help feeling that this'—he tapped the newspaper he was holding—'is very hopeful. It wouldn't surprise me a bit if you heard some news about Crabbie before long.'

His optimism seemed to be justified when, three weeks later, a stranger rang the villa at Biot and asked to speak to Patricia Rose.

James called his sister. 'It's for you, Pat,' he said. 'Some man or other. Won't give his name.'

She took the receiver and a voice asked: 'Patricia Rose?'

'Yes.'

'I have a message for you. Commander Crabb is alive. He is safe and wants you to know that you will see him in ten years' time.'

'Who are you? Is this a joke?' Pat asked.

'No joke. Commander Crabb was taken to Moscow and jailed at . . . prison' (Pat could not make out the name) 'because he refused to disclose anything about himself or his work. Colonel . . .' (again Pat could not understand) 'offered him a position in the Red Navy at a salary of £1,000 a year and assured him that he would not be asked to work against Britain. The commander agreed to join the Red Navy for the proposed term of ten years, after which time he was assured he would be free to go wherever he liked. He wants you to know that you will then be reunited with him.'

'Who *is* this?' Pat asked again.

There was a click and the line went dead.

She stood there, holding the receiver, staring blankly at her brother.

'What's the matter?' he said. 'You look as if you had seen a ghost.'

'Someone just told me that Crabbie is alive and safe in Russia and that I'll see him in ten years' time. Then he hung up on me.'

'But that's wonderful,' he said. 'Remember I told you that you'd be getting news of Crabbie before long?'

'Of course it's wonderful . . . if it isn't a hoax.'

'I don't think it is. First your pictures in that Communist paper, and now this message. It certainly looks as though Crabbie's trying to get through to you.'

Two days after the mysterious telephone call French and other Continental newspapers carried the following story:

'A French left-wing politician, recently returned from the Soviet capital, said that Commander Crabb had been taken to Moscow after being arrested by Soviet frogmen in Portsmouth Harbour on April 19.

'On his arrival in Moscow, Commander Crabb was taken to Lefortovo Prison. His prison number was 147. At first he refused to give the Russians an account of his work but after some days "became very open." He admitted investigating the automatic steering mechanism which enabled the Soviet vessel to avoid shallows and obstacles.

'A Russian official, Colonel Miaskoff, offered Commander Crabb employment with the Red Fleet at a salary of £1,000 a year. He was assured by Admiral Trobuz, head of the naval counter-espionage service, that he would not be asked to work against Britain.'

'Here, Pat,' James said, handing her a copy of a French paper containing the story, 'here's confirmation of what your mystery caller told you over the 'phone.'

'Let's pray that it's true . . . that Crabbie is alive,' she replied.

Five and a half weeks later, on 7 August 1956, *Reuter* cabled the following message from Copenhagen:

'Lionel Crabb, the frogman who went missing when the Soviet cruiser *Ordzhonikidze* brought B and K to Portsmouth in April, is in Russia, members of the anti-Communist movement *NTS* said today.

'The *NTS* members said they learned from Soviet sailors from the *Ordzhonikidze* and two destroyers which left Copenhagen today after an official visit that on the cruiser's return voyage from England part of the ship's hospital was strongly blocked off. There was someone in the area, but none of the crew knew who it was.'

When Patricia read this news item she agreed with her brother that her hope of seeing Crabb again in ten years might, after all, be well-founded.

* * *

Meanwhile, Commander Lionel Crabb was assuming more and more his new image of Lev Lvovich Korablov. He underwent continuous indoctrination with Soviet ideology while he worked hard at mastering the intricacies of the Russian language.

On Friday, 17 August 1956, he was driven from the State Security Committee rehabilitation centre at Selskoye to Khimky Naval Intelligence Station and faced Colonel Myaskov in the interrogation room.

'Your conduct sheet, Lev Lvovich Korablov, records that you have been making progress and that you understand and now speak sufficient Russian to be able to follow your occupation,' the colonel said. 'As there is no point in wasting any more valuable time, I am now transferring you to the Naval Command, where they will grade you according to your qualifications.

'I hope you realise that I am taking a personal risk in not having you passed on to the Military Tribunal for sentence but, instead, am giving you a chance to follow your occupation. I trust that you will not let me down and that you will not repay my humanitarian deed with possibly unforgivable attempts to double-cross us in one way or another.

'I can go only by the reports which I have regularly received about you. Unfortunately I cannot read what is going on in your mind. But I hope that I am not making a grave mistake by trusting you and so helping you to live as a free man, with the possibility of working honestly and to the best of your abilities in your own profession.'

For some minutes he paced the room in silence. Then he continued :

'To avoid any misunderstanding, I must tell you again that though you will become from now on a member of the Red Navy, you will nevertheless have no chance of planning any escapades or submitting misleading information about under-water examinations which you may be ordered to carry out. Every move you make while in the Red Navy will be skilfully watched and any wrong step will be punished without mercy.

'But I hope that my words of warning are unnecessary

and that, instead, you will work to the best of your ability and avail yourself of the possibility of climbing back gradually to the position you held in the British Navy. In our country we reward initiative and good work, and those who deserve it can—in a comparatively short time—become high-ranking officers.

'Have you decided to become a worthy member of the Red Navy and to work honestly and to the best of your abilities?'

'I have,' Crabb said.

'Then all that remains is for you to fill in the application form to join the Red Navy and sign it in your name of Lev Lvovich Korablov.'

Myaskov handed Crabb the official, printed paper. He filled it in, signed it and handed it back.

'Thank you,' the colonel said. 'You will now be paid eight hundred roubles, to which you are entitled for the time of your studies. I am also allowing you to select a complete civilian outfit for yourself so that you will arrive at the Naval Training Command like any other recruit. You will feel better when you finally take off your dungarees—and with them you will strip off your unpleasant past.

'As the train for Leningrad does not leave for some hours, I am allowing you to go sightseeing in our wonderful capital.'

With that Crabb was dismissed and handed over to Captain L. B. Chuikov, who showed him around Moscow, accompanied him on his train journey to Leningrad and next morning, Saturday, 18 August, handed him over to the reception officer at the Naval Training Command at Kronstadt.

* * *

Captain Nikolai Petrovich Khorneichuk, commandant of the naval training station, submitted regular reports on Lev Lvovich Korablov to Investigation Judge Colonel Aleksei Feofilovich Myaskov. They all confirmed that the new recruit continued to give satisfaction in every respect and that he was one of the best men in the command.

On Sunday, 28 October 1956, Lev Lvovich Korablov was promoted to the rank of petty officer and transferred to Operational Command, Arkhangelsk. Chief Security Officer Commander Yuriy Pavlovich Aremashvili was so impressed after Crabb had volunteered to assist an icebreaker in trouble that he recommended his promotion to second-lieutenant. Myaskov agreed and advised the appropriate department accordingly. So, on Friday, 28 December 1956, Commander Yuriy Pavlovich Aremashvili informed Crabb, or Korablov, that 'due to satisfactory conduct and special services rendered,' he had been promoted to commissioned rank and that his promotion to second-lieutenant was contained in the orders of the day.

Crabb stood speechless for a few seconds. Then he said : 'Thank you, tovarich commander. Thank you very much indeed.'

Three months later, when Second-Lieutenant Korablov's promotion to lieutenant and his posting to Operational Command Baltiysk was made known in the orders of the day, almost everyone at Arkhangelsk expressed regret that the well-liked frogman was leaving. On Sunday, 31 March 1957, Lieutenant Korablov left for Baltiysk and immediately took up his duties.

On Sunday, 5 May 1957, the news information room at Khimky Naval Intelligence Station sent a message to Colonel Myaskov. It informed him that a London Newspaper* had reported the following statement from a senior Whitehall official :

'We are satisfied that Commander Crabb did not die when he went into the water at Portsmouth near the Russian ships, here for the visit of Marshal Bulganin and Mr. Khrushchev. We have good reason to believe that he was taken aboard one of the ships and is now

* *Reynolds News.*

held in Russia.'

The message also quoted a dispatch from the *TASS* correspondent in London :

'The sudden statement of the Whitehall spokesman, published in the British Press, indicates that the British government has not closed the "Commander Crabb case" but that further action may be considered at any time.'

When Myaskov returned to his office for duty on Monday, 6 May, and read the message, he immediately telephoned Major B. M. Smirnov in Leningrad. He considered it imperative, he told him, that the prepared body should be transported to England at the earliest possible moment to end once and for all the rumours that Crabb was alive and in the U.S.S.R.

Smirnov promised to have the body examined at once by experts. He telephoned the investigation judge that same evening to advise him that the body was ready and that he was sending the doctor's report, which stated :

'At 15.52 hours examined a body in rubber frogman suit and found it decomposed to such an extent that it is most unlikely that any expert would be able to establish the cause of death. I could not, of course, carry out a full post-mortem examination, as Major B. M. Smirnov, who was present, insisted that the frogman's suit must not be opened. But my conclusion about the decomposition of the body has arisen from the fact that after the most thorough examination I was unable to state what had caused the head and both hands to be detached from the body and the arms. I am satisfied that no pathologist in the world will be able to establish the cause of death; also that no one can identify the body. The only identifiable parts are the frogman's suit and the flippers, which are still firmly on.

'It appears, however, that the lower part of the body is better preserved, though this I could not establish for certain as I was not allowed to strip it. My theory is based on the following two facts: when examining the body below the waist, I was able to feel its shape through the suit; the second point for my assumption is that below the waistband the body is sealed by a rubber band which did not permit the seawater fully to penetrate into the frogman's suit and thus decompose the lower part of the body. But though I assume that the part of the body from the waist downwards is in a comparatively good state of preservation, nevertheless I rule out the possibility of its providing sufficient evidence for identification purposes.'

Colonel Myaskov discussed the question of the 'return' of the body with Major-General Maksim Mikhailovich Chuganov, who agreed that it was essential to take the body to England as soon as possible but opposed the idea of using a Soviet merchant ship passing the English coast. He pointed out that any possible gossip about the matter by Soviet seamen in foreign ports must be prevented at the outset. Myaskov agreed and it was decided that the task should be carried out by a commander of the Red Navy.

Everything was in favour of Colonel Myaskov's operation. On Saturday, 1 June 1957, three submarines were to leave their base for Egypt. Myaskov arranged with Chief Security Officer Captain Anatol Grigoryevich Muralov to transport the body to its destination by means of the submarine flagship.

At 02.14 hours on Saturday, 1 June, the body was brought alongside the vessel. Muralov was now in full and sole charge of the top-secret mission. Under his supervision the body was placed in a steel container filled with seawater and secured by steel hawsers to be towed at a safe distance behind the submarine.

At 23.40 hours on Wednesday, 5 June, submarine headquarters received a coded radio message from Captain Muralov aboard the flagship:

'According to our detailed charts we are approaching the spot where the tide is most likely to wash charge ashore near Portsmouth.'

Forty-six minutes later another coded message was received :

'Cargo dispatched 00.59 hours. Charts and currents indicate that arrival can be expected in 48 hours.'

* * *

Just at a time when Patricia Rose's spirits were beginning to sink during those lovely sunbathed days at the beginning of June, 1957, she received another injection of confidence. Another anonymous caller rang her at her brother's villa in Biot. 'Crabb wants you to know that he is well and is longing to see you again in nine years' time,' he said.

'How can I be sure that you are talking for Crabbie?' Pat asked.

'He told me to remind you of your train journey to Portsmouth, when he looked at some photographs of you and said that you would be married soon. He said that must convince you that his message was genuine because there was no one but yourselves in the compartment.'

'Yes. The message must be genuine,' she exclaimed. 'You couldn't possibly have known about the train journey.'

'Crabb hopes that all your doubts that he is alive are now dispelled and that you won't worry about him any longer,' the caller said. Before she could ask more, he had replaced his receiver.

8

Question of Identity

On Sunday, 9 June 1957, Mr. John Randall, a 40-year-old former merchant seaman was out fishing when he saw a body clad in a frogman suit floating off Pilsey Island, a sandbank at the mouth of Chichester Harbour, near Thorney Aerodrome. With the help of two companions he hauled the headless and handless body into the boat and took it back to the mainland. There police and R.A.F. officers took over and the body was transferred to Chichester mortuary.

Dr. Donald Plimsoll King, consultant pathologist to the Chichester Hospitals Group, spent sixty-five minutes examining the remains and prepared a three hundred word report for the inquest which was fixed for Tuesday, 11 June. He found no clues to the cause of death and nothing on the well-preserved frogman suit to help with identification. 'I examined it most carefully but I was unable to trace any Navy serial numbers,' he said.

His report might be summarised: 'Unable to establish cause of death. No easy means of identification available.'

Concerning the feet, he stated: 'I feel a bit doubtful about any deformity. The state of the toes is not sufficiently abnormal to make any confident statement. There is no gross deformity.' The importance of this statement will become clear later.

Dr. King referred to the upper part of the body, held in the skin-tight rubber suit, as 'just a collection of bones,' with the hands missing; but he described the lower part, held by a thick rubber band around the waist, as in 'quite a good state of preservation'.

On Monday, 10 June, Mr. G. F. L. Bridgman, the Chichester coroner, said: 'I propose to open the inquest tomorrow afternoon and adjourn for a fortnight. Whether the final inquest will be held in public or in camera I do not know.' He said he had received the report of Dr. Donald Plimsoll King, the pathologist who had made the post-mortem examination, and added: 'Quite honestly, we know little more than we did before. There is no question at the moment of having established identification. This is not a straightforward case where a person who was with someone at the time can be called to say: "That is the man." We have appealed to the Admiralty to help us and they have agreed to do so. They will be sending experts tomorrow to see the body and the clothing.'

Mrs. Margaret Elaine Crabb, whose marriage to the commander had been dissolved almost three years earlier, saw the body at Bognor Regis mortuary, to which it had been transferred from Chichester because it could be better preserved there. She did not identify it. She told the police that her former husband's feet were readily identifiable because he had 'buckled' toes. She pointed out that the toes on the body at which she was looking were *not* deformed.

Although the identity of the corpse found floating off Pilsey Island had not yet been established, Superintendent S. L. Simmonds, of Chichester police, stated that there must be a strong assumption that it was that of Commander Lionel Crabb, R.N.V.R., who had disappeared after trials with underwater apparatus near Portsmouth on 19 April 1956. He said that the body was headless, was without hands, and quite obviously had been in the water for some considerable time. From the action of the currents it could be assumed that if Commander Crabb had been drowned off Portsmouth, the body would have been carried in the direction of Chichester Harbour, where it was recovered.

'This is not just an ordinary body,' Chichester police

spokesman said. 'There are security, political and other reasons for precaution.'

On Tuesday, 11 June, at Chichester magistrates court, Mr. Bridgman opened the inquest on 'an unidentified man'. The only evidence given before the hearing was adjourned until Wednesday, 26 June, was that of the coroner's officer, P.C. D. Castleden. Mr. Bridgman put eight questions to him :

'This morning did you go to the public mortuary here with me?'

'Yes, sir.'

'Did you view the remains of the body of a man?'

'I did.'

'And certain articles of clothing?'

'I did.'

'Did they consist of a black rubber frogman's suit and flippers?'

'Yes, sir.'

'A pair of maroon-coloured bathing trunks?'

'Yes, sir.'

'A pair of blue socks?'

'Yes, sir.'

'A pair of fawn-coloured combinations?'

'Yes, sir.'

'And a pair of blue woollen combinations?'

'Yes, sir.'

The proceedings lasted forty-five seconds.

On Wednesday, 12 June, under the headline MRS. CRABB UNABLE TO IDENTIFY FROGMAN, the *Daily Telegraph* reported :

'Mrs. Margaret Crabb, former wife of Commander Lionel Crabb, the naval frogman, failed yesterday to identify the body found off Pilsey Island, Chichester, on Sunday, as that of her ex-husband.

'Inquiries in Chichester show that some of those in charge of the investigation have received secret briefing on the circumstances relating to Commander Crabb's fate. For this reason it is unlikely the full facts about his mission will ever be made public. Cause of death will probably never be known because of the decomposition.

'The failure of Mrs. Crabb to make any identification means that, apart from the clothing, the burden will rest on scientific tests and measurements. Unless the clothing is positively identified or fresh evidence is forthcoming, the body may never be recognised by the authorities as that of Commander Crabb.

'Mr. Bridgman said last night: "It may be some time before I issue a certificate for the burial of the body. I will not issue a certification while identification remains possible."

'The naval officer who examined the frogman's equipment at Chichester police station yesterday is Lieutenant W. Y. McLanachan, a torpedo and diving expert from *H.M.S. Vernon*, Portsmouth. He has been placed at the disposal of the West Sussex Constabulary.

'The fact that Lieutenant McLanachan was not at the inquest and that only underclothes, a frogman's suit and flippers were identified, suggests that some other secret equipment might have been found on the body.

'Also, the mortuary keeper at Chichester was told not to carry out his usual procedure of preparing the body for the post-mortem examination. This would suggest that the police, at the request of the naval authorities, might have removed something.'

* * *

The mystery deepened when it was disclosed that three days before the body had been washed up near Pilsey Island, three Soviet submarines had passed through the English Channel. Not only in Portsmouth and throughout the United Kingdom but in countries overseas new rumours circulated and old ones were revived and hotly debated. Most widespread were two theories considered by many to be nearest the real truth about the frogman's disappearance and the sudden finding of the decomposed, unnamed body.

The first was that while attempting to examine the hull of the *Ordzhonikidze* Commander Crabb had been captured by the Russians, taken to the U.S.S.R. and interrogated. When he had refused to tell the Russians what they wanted to know, they had killed him. One of the three Soviet submarines which had passed along the English coast three days before John Randall made his gruesome 'catch,' had carried Crabb's body and had released it in such a way that the currents would float it towards Portsmouth or Chichester, where it would be identified.

The second theory held that the body was not that of Commander Crabb and that, for one reason or another, the submarines had been used to plant on the British authorities a 'man who never was,' in order to put an end to the long speculation on the frogman's fate. The real Crabb, said this body of theorists, was probably alive in a Russian prison.

* * *

The official investigations into the mystery came to an end when the inquest was resumed on Wednesday, 26 June, at Chichester.

Dr. King, who had first examined the headless and handless corpse, told the coroner that the dead man's feet measured 8¾ inches—small for a man—and had distorted big toes. There were signs of a scar on the left knee. Dr. King said he had formed the opinion that the body was that of a man about 5 feet 6 inches in height. His legs were muscular and straight and, apart from the toes, there was no other deformation. He estimated that the body had been in the sea for at least six months and might well have been immersed for fourteen months.

Mr. John Randall, of Bosham, Hants, gave evidence of finding the body in the sea. He said he had been out fishing with two other men when he saw an object floating in the water. On reaching it he realised that it was the body of a man in a frogman's suit.

Details of Commander Crabb's service career were given by Mr. Gordon William Bostock, a temporary clerical officer in the Admiralty Reserves Department and the only naval witness. He said that in 1952 Crabb was promoted to the rank of Lieutenant-commander and that in April, 1955, he was finally released from naval service. Since that date he had never been employed in the service of the Royal Navy or had done any training.*

Mr. Sydney James Knowles, B.E.M., gave details of his service with Commander Crabb during the second world war. He said that he joined the Navy in December, 1939, and two years later met Crabb and did underwater work with him. Explaining how Crabb got a scar on his left knee, Mr. Knowles said that in 1945 they were in Leghorn, Italy, together. At that time captains of ships were instructed to hang rolls of barbed wire below the waterline of their vessels to protect them from attack by Italian frogmen. One morning he and Crabb had dived to find limpet mines attached to the hull of an American ship.

'We had to swim under it,' he said. 'While we were going down, a tug passing overhead cast a wash, throwing both of us against the barbed wire. When we got back I noticed that Commander Crabb had a wound on his left knee and I dressed it. Three weeks later, when I was working with him again, I noticed a scar in the shape of an inverted Y about the size of a shilling piece on the left side of his knee.'

After the war, he said, he had seen Crabb from time to time. Crabb then had worn a two-piece rubber suit which differed from others because it had a neck seal

* *The Navy List 1956*, published by *Her Majesty's Stationery Office*, however, lists on page 225 under the heading *Royal Navy Volunteer Reserve* the following entry: 'Crabb, L. K. P., O.B.E., G.M. Cr (Sp) L 11 30 June 52,' thus confirming that Commander Crabb was still a serving R.N.V.R. officer.

instead of a hood.

Miss Amy Thomas, manageress of service flatlets in Hans Road, said Crabb had occupied one for five years until April, 1956. Early that month he had said he was going away for a few days. He left on April 17. He never returned. While living at the flatlets he had been engaged upon some business. She thought he had been working in the furniture trade.

Crabb's former wife, Mrs. Margaret Elaine Crabb, said that she and Crabb had been married on 15 March 1952. They had lived together until April 1953, when she began proceedings for divorce. She obtained a decree by an undefended petition which was made absolute in December 1954. During their married life Crabb was serving as a commander in the R.N.V.R.

Mrs. Crabb confirmed details of her former husband's height and other points given by Dr. King. She agreed that Crabb was short, had straight muscular legs and what she described as 'hammer' toes. She thought he took size six shoes. When she went to identify the body she could not say from looking at the feet whether it was the body of her former husband—but equally she could not say that it was not.

Mr. Eric Blake, managing director of Heinke & Company, of Bermondsey, manufacturers of underwater swimming suits, described suits which he had supplied to Commander Crabb. He said he had supplied Crabb with three frogman suits, the last in October, 1955. It was unusual because it had a neck seal without a hood. He had been shown the outfit recovered from the sea and found it identical with the type he had sold to Commander Crabb.

Recording an open verdict, the coroner said that it was impossible to determine the cause of death. He continued: 'We have all been warned from time to time in the legal profession about a chain being as strong as its weakest link, but there is also such a thing as a number of incidents which are minor indications building up to a conclusion which I do not think can be resisted.

'I think it would be beyond all our ideas of possible coincidence if all these different things—the size of the feet, the scar, the colour of hair, the sort of legs, the supply of identical suit—if all these things were to be put down as sheer coincidence.

'Looking at the evidence in this case I am quite satisfied that the remains which were found in Chichester Harbour on 9 June were those of Commander Crabb, who was last seen, by a witness called today, on 17 April 1956, in London.'

The inquest lasted less than an hour. No evidence was given by anyone about the last two days of Crabb's life—18 and 19 April 1956.

After the inquest Mr. Alastair W. Geddes, solicitor for Commander Crabb's mother, Mrs. Beatrice Crabb, issued the following statement:

'There have been irresponsible conjectures regarding the last operation upon which Commander Crabb was engaged prior to his disappearance, including even suggestions that he was in the employ of a foreign power. It is desired therefore to emphasize that every particle of evidence in our possession points conclusively to the fact that this very gallant gentleman died, as he had lived, in the service of our own country and of no other.'

Asked if he would elaborate, Mr. Geddes replied:

'I cannot say more. Our hands are tied.'

On the evening of 26 June, an Admiralty spokesman said:

'The Admiralty can add nothing to its original statement or to the statement of the Prime Minister.'

* * *

Despite Mr. Bridgman's dissertation on 'a chain of coincidences,' it was commonly felt that the inquest had not thrown full light on the problem of whether the body found floating in Chichester Harbour were really that of Commander Crabb. More and more people began to speculate and in some cases express the firm opinion that the coroner's findings might have been the outcome of instructions received from 'the highest quarters'.

The main questions which puzzled the public were:

How was it that neither the pathologist nor Crabb's former wife were able to find any deformation of the toes when first examining the body—and what made the pathologist admit, at the resumed inquest, that the toes of the body were deformed?

How was it that after a careful examination, lasting sixty-five minutes, the pathologist had not found a scar on the body's left knee—and what made him admit, at the resumed inquest, the presence of such a scar?

Did the pathologist disclose his real findings on the examination of the body in his three hundred word report *before* he received orders to confirm such findings as were required for a declaration that the body was that of Commander Crabb?

Did it occur neither to the pathologist nor to the coroner that it was more than strange that the head and *both* hands—all the vital parts which could have provided conclusive identification—were missing from the body?

Why was this question not fully explored during the inquest, especially since it was known that Soviet submarines had passed through the Channel three days before the appearance of the body and that Crabb had vanished while diving in the vicinity of the Russian cruiser *Ordzhonikidze*?

Why was Lieutenant McLanachan, who had examined the equipment, not called to give evidence at the inquest—and why was 'Mr. Matthew Smith' not called?

Mr. Rankin, Labour M.P. for Govan, Glasgow, doubtless had these and other questions in mind when in Parliament, on Thursday, 4 July, 1957, he asked the

Prime Minister whether he could now make a statement in the circumstances in which Commander Crabb had disappeared in Portsmouth Harbour.

Mr. Macmillan replied: 'I have no statement to make on this subject.'

Mr. Rankin then asked: 'Is the Prime Minister aware that after the inquest the solicitor representing the Crabb family stated the Commander Crabb had died in the service of his country? Does the Prime Minister accept that view; and if he does, will he see that the appropriate pension is issued to the widow in addition to the solitary *ex gratia* payment of £10 which so far has been made?'

Mr. Macmillan said that he would look into the second part of the question. As for the first part, he could make no further statement. He must stand by the statement made by Sir Anthony Eden in the debate on 14 May 1956.

* * *

On Friday, 5 July 1957, the body of the 'unknown man' was at last buried at Milton Cemetery, Portsmouth. There were no official representatives of the Royal Navy at the funeral.

Officially, the last chapter in 'The Commander Crabb Affair' had been written.

9

Stranger in a Train

Patricia Rose attended neither the inquest at Chichester nor the funeral at Milton Cemetery, Portsmouth. She remained at her brother's villa in Biot and followed the proceedings in England through the British and Continental newspaper reports and radio broadcasts. The more details which were published about the decomposed, nameless body, the more confident she became that it was not that of Commander Crabb.

One day her brother read aloud the report from the Chichester correspondent of a Paris newspaper concerning the conflicting evidence which Dr. King had given in his original three hundred word statement and at the resumed inquest.

'But Crabbie had deformed toes,' Pat exclaimed. 'You remember how odd they were, James. Nobody could possibly make a mistake about that.'

'I agree—but apparently the pathologist was none too certain.'

'What does it say there about Crabbie's ex-wife?'

'Oh! she was taken to the mortuary at Bognor Regis to identify the body, but it was a flop as far as the authorities were concerned. She told the police that the body was not Crabbie's because the big toes were normal.'

'Well, she ought to know, if anybody does. It can't be Crabbie, so those 'phone messages we have been getting could be genuine.'

'I was always inclined to believe that they were,' James said. 'There seems to be little doubt that the Admiralty used the inquest to put an end to all the rumours. Reading between the lines, it's not difficult to see that—and this sort of thing has happened in other countries. The powers that be probably hope that public interest will wane and the whole business will soon be forgotten.'

* * *

If the authorities really had any such hope, they were quickly to be disabused. British and foreign newspapers and magazines persisted in discovering and publishing new developments. Reports from Scandinavia, France, the Middle East and other countries disclosed that Red seamen who had been at Portsmouth during the B and K visit confirmed on various occasions that on 19 April 1956, a British frogman had been captured by Russian divers and that he had been imprisoned in the *Ordzhonikidze's* hospital quarters and smuggled out of Britain. Because similar stories were told at different times in different ports, by different individuals—each such disclosure and informer being unconnected with the others—the belief persisted that the body found off Pilsey Island was not that of Crabb and that the whole affair was an ingenious Soviet 'plant'.

When the sea washed up a bottle containing a message obviously written by Crabb, everything was done to prevent newspapermen from getting the full details concerning the incident. As far as officialdom was concerned, Commander Crabb was dead and buried and they were determined to let things stay that way.

Mrs. Beatrice Crabb immediately recognised the handwriting as that of her son. Handwriting experts believed that the writing was Crabb's but played safe by admitting that a clever forgery could not be ruled out. Other experts claimed that the message, even if it were genuinely from Crabb, proved nothing. They pointed out that it was undated. Crabb could have written it and thrown it into the sea long before he disappeared.

The bottle was corked perfectly, the paper and writing were unaffected by seawater and there were no marks on the glass which might point to the bottle's origin, they said. They did not explain why the commander should have indulged in such a childish game without reason.

Mrs. Beatrice Crabb dismissed their views. She said categorically: 'Their idea that it could be a clever forgery is ridiculous. Why should anyone wish to forge it? That's preposterous! And the insinuation that Lionel could have written it long before he disappeared is equally nonsensical. If he had done it for a prank, to see how long it would take before the bottle was picked up somewhere, he would have written a different message and would have made sure it bore his full name and address, so that he could get it if it were found. He wouldn't have just signed it with a capital "L". No! All the experts' "ifs" and "buts" are rubbish. I am quite satisfied that Lionel intended to let me know in some indirect way that he is all right.'

* * *

Two years after Crabb's disappearance, Patricia Rose returned to London. She told mutual friends about the messages she had received while in the South of France and was glad to find that few believed the commander to be dead. When she had talked with her prospective mother-in-law and had heard from her all about the bottle-message, she was as convinced as the old lady that Crabb was doing all he could to let them know that he was well and thinking of them.

Five months after her return the telephone in her Knightsbridge apartment rang at 6.30 p.m. on three successive evenings. On the first occasion a strange voice said:

'I am 'phoning you at Crabbie's request' and told Pat that he had met him in the Crimea. With that he hung up.

During the second call the stranger told Pat that Crabbie had asked him to tell her about a secret drawer in his desk, in which he used to keep money. He described the drawer in such detail that Pat knew that only Crabb himself could have given the man such precise information.

At the end of his third and last call, the voice said: 'If you don't have to, don't talk about these telephone conversations. Let people think what they wish.' Once more he replaced the receiver without revealing his identity.

Pat left London and moved to Sussex; but she did not neglect her friends and frequently travelled to town to see them. One evening, as she boarded the train at Victoria after such a visit, a stranger walked into her compartment and asked: 'You are Patricia Rose, aren't you?'

'Yes,' she replied. 'How do you know?'

'Never mind. . . . You know Commander Crabb is alive.'

'I think he is,' she admitted.

'He *definitely* is,' the man corrected her. 'I can prove to you that he was brainwashed in Moscow and I know for a fact some of the things he said.'

'Who are you? Why have you tracked me to tell me all this?'

'I want you to be certain that your fiancé is alive in Russia and longing to be reunited with you when his tour of duty with the Red Navy terminates in August, 1966. That's all.' He smiled. 'You don't trust me, do you? Well, that's understandable.'

'I don't know what to think,' Pat said uncertainly.

'Did Commander Crabb have a saying of "Poppycock" when he got angry or irritated?'

'Yes, he did. But that's no proof that you know whether he is alive in Russia. One of his friends could have told you. He's well enough known. You could have found out all sorts of things about him——'

'Of course I could have done, but I did not,' he interrupted. 'I have never met Commander Crabb and I know none of his friends.'

He then disclosed that he belonged to an anti-Communist underground movement which had sent him to give her the full facts. He gave personal details about Crabb and herself which convinced her that his informaton must be authentic.

'Who are you?' she asked again.

'I'm afraid I can't tell you that,' he said, 'but you can rest assured that everything I have told you about Commander Crabb is true and that you will be seeing him again.'

When the train stopped at East Croydon he left the compartment and hurried down the staircase that led from the platform.

Three months later, as she left the train on another London visit, an old gentleman with a full beard stopped her near the platform barrier and showed her a photograph. 'This is Commander Crabb,' he said abruptly.

The picture was slightly out of focus and Pat wanted to examine it carefully. 'May I have it, please?' she asked.

'I'm sorry,' the man said. 'I was only asked to show it to you so that you would know Commander Crabb was alive and that you would meet him again. I am not authorised to let you keep the picture.' She noticed that he spoke with a foreign accent, perhaps mid-European.

'How did you recognise me, and how did you know that I should be coming to town on this particular train?' she wanted to know.

'You ask too many questions.' He took the photograph, put it into his briefcase and, without another word, turned and walked away into the crowd.

She looked after him, perplexed. He had given her no opportunity to examine the photograph closely. Was he genuine or some kind of confidence trickster? How had he known that she would be in London that day? She could find no answer that would satisfy her.

Still another message reached her in October 1959. This time, though the caller would not give his name, there could be no doubt from the voice that he was German.

'Is that Patricia Rose, Commander Crabb's fiancée?' he began.

When Patricia confirmed that he was speaking to the right person he went on:

'I have good news from Commander Crabb for you. He is in good heatlh and is quite happy. He is now in the Vladivostok Naval Command and his rank is first-lieutenant.'

'How do you know all this?'

'It has come through the East German underground group. East Germany is Communist Germany—Russian-controlled—behind the Iron Curtain, you know. They get very reliable information from Russia and we know that our informers are very reliable.'

'Who are you?'

'I cannot tell you. But I can tell you that I work with a refugee organisation and that we get very reliable information from East Germany. You can believe me. I have no intentions to tell you something which is not true.'

Pat said: 'I didn't think you were playing games with me, but it's only natural that I should like to know something about you so that I can make up my mind whether to believe you.'

'So! I cannot force you to believe what I told you but it is the honest truth. All the information we get through the East German underground group is always absolutely correct.'

'How did you find me?' Pat was anxious to know how all her mysterious callers knew where to locate her.

'I got your telephone number from friends. You had

FIG I Commander Lionel Crabb. This photograph was taken three years before the Commander's disappearance and the snapshot comes from the family album.

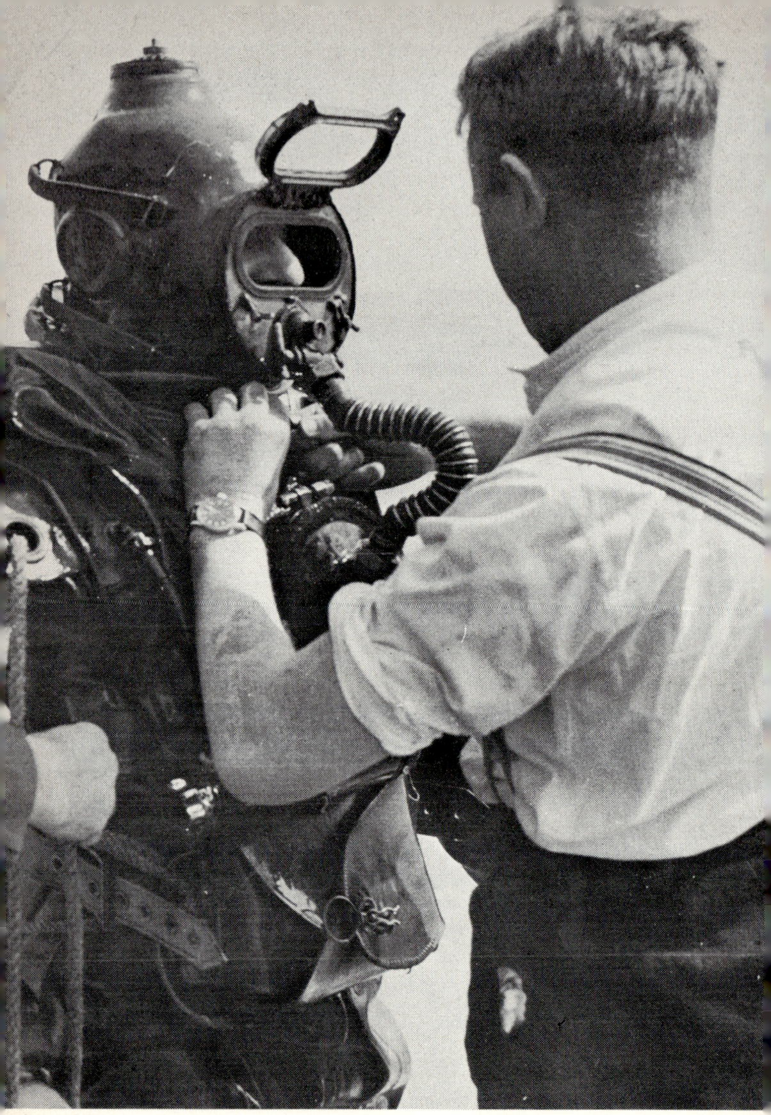

FIG 2 Commander Lionel Crabb in diving gear. This photograph was taken in England in 1950.

FIG 3 Commander Lionel Crabb alias First-Lieutenant Lev Lvovich Korablov (extreme left) aboard a Red Navy vessel. This photograph was obtained by members of the Russian anti-Kremlin underground movement in the autumn of 1959 and smuggled out of the U.S.S.R.

FIG 4 Commander Lionel Crabb (right) in the Red Navy. Although this picture was obviously taken at the same time as Fig 3 it was not smuggled out of the U.S.S.R. until months later. When it reached England, relatives and close friends positively identified the man on the right in the picture as the missing commander.

FIG 5 Soviet Investigation Judge Colonel Aleksei Feofilovich Myaskov, who offered Crabb employment in the Red Navy and who arranged the planting of the headless and handless body which was found at Pilsey Island.

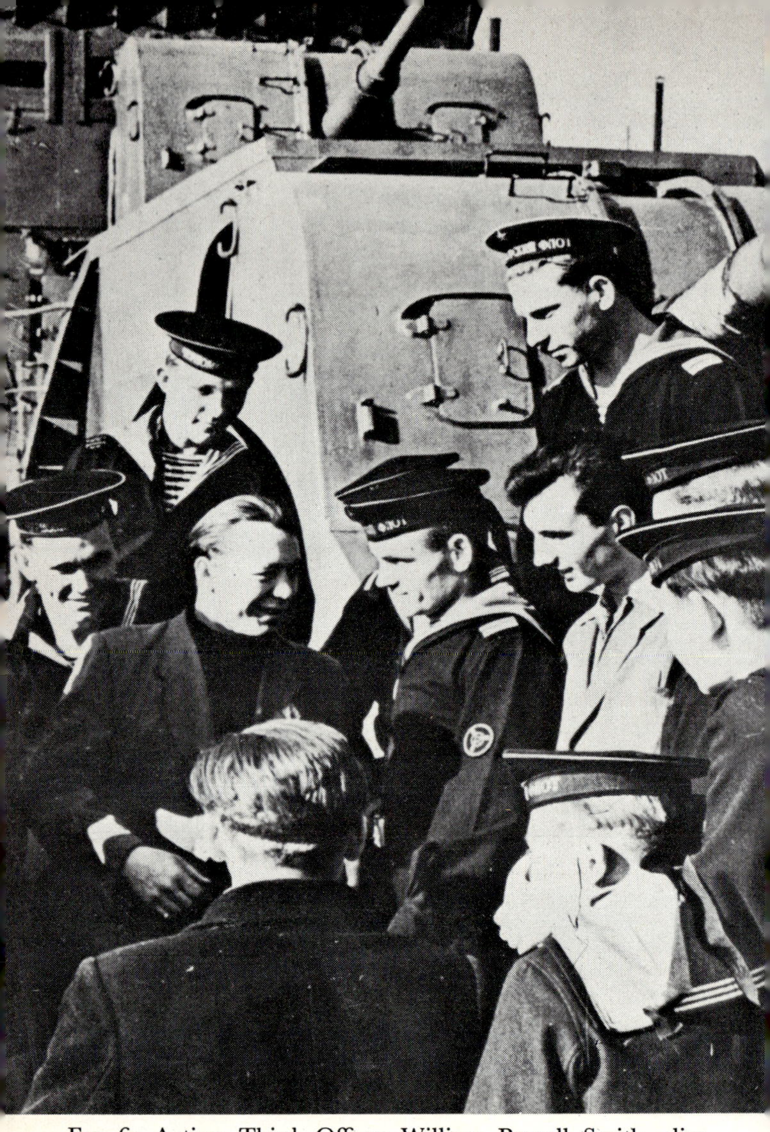

FIG 6 Acting Third Officer William Russell Smith alias Rodion Vladimirovich Smirnov (second from left) aboard a Red Navy vessel.

'It's our Russell,' said his father, Mr William Smith, after having studied the photograph most thoroughly for some considerable time.

FIG 7 Soviet cruiser *Ordzhonikidze* berthed in Portsmouth Harbour during April 1956. Alongside is the *Smotriashchin* and another destroyer of the Soviet Baltic Sea Fleet.

Fig 8 Patricia Rose, Crabb's fiancée, at the swimming pool of her brother's villa in the South of France. This is the photograph, taken by Matthew Smith from the Commander's wallet, which was published in Iron Curtain newspapers.

several messages from Commander Crabb before. I am sorry that I cannot give you more details.'

'Where did you say Crabbie is now?'

'In Vladivostok, with the Red Navy. He is a first-lieutenant and his Russian name is . . .' (Pat thought it sounded like 'Komarov' or 'Koravov'.)

'Couldn't we meet somewhere, so that you could tell me more?' she suggested.

'I am sorry. That is impossible. I am just leaving England. I have told you what the message for you says. The only thing I have not told you is that your fiancé is longing to see you again when his contract with the Red Navy expires in seven years' time. That is all the information I have for you.'

'Would it be possible for me to send Crabbie a message through your German underground people?' Pat asked.

Her question remained unanswered for the man at the other end of the line disconnected the call.

'It is always the same,' she complained to her aunt. 'Every time I get a message from Crabbie and try to find out more about it, they hang up on me.'

Nevertheless she was delighted to have had further news and was grateful to the anonymous people who took the trouble to contact her.

10

The Smuggled Dossier

In November 1959, full light was thrown at last upon the riddle which had caused such great public interest in Britain and most other countries of the civilised world. The revelation came in form of a Soviet secret *dossier* which was smuggled from behind the Iron Curtain and reached England quite unexpectedly—almost, indeed, by accident.

I was working at that time for the *Empire News,* a national Sunday newspaper to which I was able frequently to give important inside information which I received from contacts in the U.S.S.R. and her satellite countries. There were many occasions when I revealed what had taken place at top-level meetings in the Kremlin, together with other important happenings in the Communist world, weeks before the news was confirmed and published in the world Press. As a result, my editor asked me to concentrate on finding out what had happened to certain Britons who were thought to be, voluntarily or against their will, somewhere behind the Iron Curtain.

In due course a great deal of information about missing British civilians and servicemen was smuggled out by my friends in Eastern Europe and in all cases proved to be accurate and precise. Nevertheless I was stunned when I extracted from the package which arrived in November, 1959, a Soviet State Security *dossier* on Commander Lionel Crabb. The accompanying note from my contacts said that it had been sent in answer to my request for further information about missing Britons. They hoped that this account of the fate of the British frogman might be of interest. Despite the extensive and continuous Press coverage in Western countries of 'The Commander Crabb Affair,' little news had been allowed to penetrate the Iron Curtain and consequently my informants were unaware of the vital importance of this top-secret document.

Having had ample proof in the past of the reliability of my contacts in the Communist world I was fully satisfied that the dossier was absolutely genuine. So, although it had reached me more or less accidentally, I decided to translate the full text into English and publish it.

The *dossier*, headed 'The Commander Crabb Case' was unique. It contained the original reports from Captain G. F. Stiepanov, commander of the *Ordzhonikidze*; from Soviet State Security officers who had escorted Crabb to Moscow; from investigation judges, prison commandant and prison officers, interrogators and all those who had direct dealings with the British naval commander. It recorded everything which had happened from the day of the arrival of the Soviet warships in Portsmouth on 18 April 1956. A foreword stated that at the conference held in Moscow from 3 to 10 August 1959, of heads of State Security forces in all the Iron Curtain countries, it had been ruled that the secret police headquarters of the People's Democracies should receive 'translations of secret Soviet *dossiers* which could help the foreign comrades of the State Security headquarters to learn from the experience of our best-trained investigation judges how to proceed in seemingly hopeless cases, and what methods and tactics to apply to break down the prisoner without the application of drugs, violence or other such methods.'

One of these *dossiers,* which were considered to be 'useful examples of how to proceed in even the trickiest situations,' was that dealing with the case of Commander Crabb. It had been translated officially into Bulgarian, Czech, Hungarian, German, Polish and Rumanian, and in due course sent out to the Soviet chief instructors who are the big men behind the scenes in all the satellites and direct not only the Soviet puppet government leaders but also the individual State Security forces.

The *dossier* ended with a report from Captain Aleksandr Borisovich Nizhansky, Chief Security Officer, Operational Command, Baltiysk, after the decomposed, headless and handless body had been identified as that of Commander Crabb by the Chichester coroner and buried in Milton Cemetery, Portsmouth. At that time, the *dossier* revealed, Crabb was serving as Lieutenant Lev Lvovich Korablov with the Red Navy Operational Command Baltiysk. In a postscript headed simply 'L. L. Korablov,' regular monthly reports from the chief security officers of the various operational commands were recorded.

'As these reports deal only with L.L. Korablov's conduct and service,' the postscript concludes, 'it is unnecessary to quote them in full. They are of no special value to heads of State Security headquarters because no more than common knowledge could be learned from them.

'In order to show, however, that the tactics used by the brilliant brains of the Soviet security officers concerned in the case of Commander Crabb were not only ingenious examples of perfect tactics but also prove that a flexible approach to individual cases can show excellent results, it is added that Lieutenant Lev Lvovich Korablov was transferred in January, 1959, to Operational Command, Sevastopol. On the recommendation of Chief Security Officer Commander Ivan Aleksandrovich Purtsov, Lieutenant Lev Lvovich Korablov was raised to the rank of first-lieutenant on 31 August 1959. On 7 September 1959, he was transferred to the Far Eastern Naval Command, Vladivostok, because this operational command urgently requested "highly skilled frogmen-divers with vast experience".'

* * *

In February 1960, such headlines as 'Crabb Alive'—'Crabb Works For Russia'—'Crabb *Dossier* Secrets' and 'The New Mystery of Buster Crabb' appeared in British, American and other newspapers. The accompanying stories reported that my book, *Frogman Extraordinary,* then nearing completion, claimed to give the answer to the almost four-year-old mystery. In the U.S.S.R. and her satellites, however, Press and radio were ordered to maintain complete silence and to ignore stories which were being published in the West. The only exception was made in an English-language broadcast from Moscow, which said that 'anyone believing that Commander Crabb is alive in Russia must be looking through a cracked mirror'.

Reports which by-passed the Iron Curtain censors and reached England while *Frogman Extraordinary* was in the press divulged that despite the most stringent orders to all newspapers and radio networks to 'ignore the Commander Crabb affair,' the Russian secret police had put into operation immediately a most concentrated and ruthless investigation. The security bosses were determined to discover where and when the leakage had occurred.

Men and women who had direct access to secret *dossiers* of any sort were brought before the most experienced interrogators and were grilled thoroughly for days and nights on end. At last temporarily released, each was warned that he or she might be re-arrested at any time and tried for having been partner to the 'treachery'.

Thoroughout the Spring of 1960 the secret police, naval intelligence officers and other investigators continued to round up and question suspects, but despite their most intensive efforts they could not discover how, when, by whom, or from which country the *dossier* had been smuggled to England. Even when my book, containing photostats of the document, was published it did not help them much. The fact that it was a copy of the German translation which had reached me only narrowed the circle of suspects; for Moscow-made German translations had been sent not only to the State Security headquarters of the East German Democratic Republic but to many other satellite state security chiefs.

Although Colonel Myaskov and his fellow-investigators had an important lead, their task was still not easy for those behind the Iron Curtain who—often at the risk of liberty or life—supply the Western world with authentic and most important documentary evidence of what is going on in the most secret quarters of the Soviets usually have no direct access to secret files. They are fearless, outwardly unimportant men and women who somehow manage to 'borrow' a spare copy of the original document (which is seldom disturbed); and the fact that the 'borrowed' Crabb *dossier* was safely back in its place when the pre-publication excitement about *Frogman Extraordinary* began, made it impossible for the investigators to find out where, when and how the leak had occurred. Consequently, the actual culprits were not detected, but in order to frighten off others and try to prevent any similar leakage in the future, the Russian security men pounced on innocent citizens who had nothing to do with the 'abstraction'.

Neither the arrests of guiltless scapegoats nor the complete reorganisation of all State Security bureaux affected any of the group responsible for 'borrowing' the Crabb *dossier*. Furthermore, the group made it known that they had obtained a photograph of Commander Lionel Crabb, alias First-Lieutenant Lev Lvovich Korablov, aboard a Soviet warship and were doing their best to smuggle it out to me as soon as possible.

* * *

Following the news of the impending publication of

Frogman Extraordinary, on 17 February, 1960, in the House of Commons, Dame Irene Ward asked Mr. Selwyn Lloyd, then Secretary of State for Foreign Affairs, what recent information he had received regarding the whereabouts of Commander Crabb. Mr. Awbery also demanded a statement.

'No recent information about Commander Crabb has been received by Her Majesty's Government,' Mr. Selwyn Lloyd replied.

'Can my right honourable and learned Friend give us a little more assurance, having regard to some wild statements which have appeared in the Press and the disquiet which some of those observations have aroused?' Dame Irene Ward persisted. 'Has he not got any further statement to make on the subject, on whether we know definitely what happened, and whether there has been identification of the body?'

'I would remind my honourable Friend that at the inquest in June, 1957, the coroner expressed himself as quite satisfied that the remains which were found in Chichester Harbour on 9 June were those of Commander Crabb,' said Mr. Lloyd.

Mr. Awbery then asked: 'Is the Foreign Secretary aware that there is a possibility that the coroner was wrong on that occasion? Is he aware also that a book is being published shortly by publishers in London who state that they are satisfied with the authenticity of the information received from Russia? They are publishing this. Could the Foreign Secretary give us some information about the publication of this book?'

'I do not know about the book,' Mr. Lloyd replied. 'I have no other information about Commander Crabb than that which I have given to the House. I have no reason to think that there is any truth in the suggestion that the coroner was wrong.'

'Has not the time come when the Government should come clean with the House about this case, and ought not the House to know who was responsible for giving instructions to undertake that rather shady escapade?' Mr. Fernyhough asked.

'That is quite a different matter from the two questions on the Paper,' Mr. Lloyd decided—and terminated the debate.

Next day, Commander J. S. Kerans, D.S.O., M.P., sent his evidence to the Foreign Office and asked the Secretary of State for Foreign Affairs if he would make a statement on the most recent information he had concerning Commander Crabb.

Mr. R. Allen replied for Mr. Lloyd: 'I have nothing to add to my right honourable and learned Friend's reply yesterday.'

The Government was continuing with the line taken by Sir Anthony Eden, as Prime Minister, on 9 and 14 May 1956.

11

Sound and Fury

Despite the stringent measures taken by the State Security chiefs in the U.S.S.R. and Eastern Europe, additional evidence that Crabb was serving in the Red Navy continued to reach England. In April 1960, my Russian contacts succeeded even in smuggling out a photograph of Investigation Judge Aleksei Feofilovich Myaskov—a photograph unobtainable from any Soviet Press agency or newspaper archives because there was a strict rule that photographs of certain high officials of the U.S.S.R. must not be made public. A message enclosed with the print informed me that the promised picture of Commander Lionel Crabb, alias First-Lieutenant Lev Lvovich Korablov, would follow 'as soon as practicable'.

Weeks went by without further communication; then on Thursday, 26 May 1960—five days before publication of *Frogman Extraordinary*—the long awaited snapshot showing Crabb in Russian uniform aboard a Red Navy ship arrived. The accompanying note read: 'We do not know which of the men in the photograph is First-Lieutenant Lev Lvovich Korablov because none of us has ever seen him, but we are positively assured that he is in the picture. Our friends who are serving with him have given us the following description: "Short, about 166-167 centimetres, dark hair, thick eyebrows, light brown eyes (sometimes changing in colour depending on light), large nose, deep lines on face, unusual chin; if unshaved, light reddish-yellow stubble." This description should help you to establish his identity.'

Although one of the men in the picture was clearly

Commander Crabb, I decided to have the fact confirmed by competent people before releasing it. My first choice would have been Crabb's mother but I learned that she had died three months earlier. Patricia Rose was on holiday in the South of France and could not be reached. Mrs. Margaret Elaine Crabb, the missing frogman's former wife, was the only person available in England who knew Crabb really intimately. Consequently she was the first to be shown the picture—a group of four men in Red naval uniform. The man whom I had recognised as Crabb (or First-Lieutenant Korablov) was not specially marked on the print, nor did the reverse side bear any caption or other wording which could indicate the identity of any of the men or where the picture had been taken.

That evening I went with my publisher to see Mrs. Crabb at her bungalow on the Kent coast. Since she had not been told the purpose of our visit she had asked a friend, a Southern Television executive, to be present. She explained in almost hostile tone that she had to be careful when talking to journalists, writers, publishers and people of that sort and, to ensure that anything she said would not be twisted, she had arranged for an independent witness to be present. Her guarded attitude did not ease during the early stages of our interview.

In the middle of the ensuing conversation I took the print from my briefcase and handed it to her. 'Would you please look at this?' I asked.

She examined the photograph closely and in utter silence for a considerable time. After a while she showed it to her friend and whispered something. The television man nodded and replied, also inaudibly. Mrs. Crabb continued to study the photograph; then, pointing to one of the men, who was looking down at some fixture on the left of the picture, she said: 'This looks very much like him. I believe it is Crabbie.'

Mrs. Crabb had not been given an advance copy of

Frogman Extraordinary and knew nothing about the Soviet *dossier*. We left a copy of my book with her.

Later, having read it and having examined the snapshot again and again, she told a reporter on the staff of the *Empire News*: 'If this is not my former husband, Crabbie, he must have a double. Personally I believe the man is Crabbie and that he is alive and serving in the Russian Navy.

'Coupled with the tape-recordings published in Mr. Hutton's book, which read to me exactly as Crabbie would have talked in the circumstances, I feel that the evidence is overwhelming.

'I now believe that he is in Russian hands. But I am no less certain that whatever his reason for co-operating with the Russians, he would never betray Britain or consent to work against his country. It is not in the nature of the man.'

Asked why she appeared so certain that the man in the photograph was her former husband, she replied: 'I have examined carefully an enlargement of the head of the man on the left, feature by feature. The resemblance to Crabbie's face is startling. There are even dark patches on the cheekbones which could well be the twin tufts of hair which Crabbie always affected after he had removed his full beard.

'The face is thinner than Crabbie's used to be, but that is understandable. The man in the picture is smiling exactly as Crabbie used to do.

'The deep-scored lines which I remember from nose to upper lip—Crabbie called them his "lines of discontent"—are clearly visible. Crabbie had very heavy brows, an aqualine nose and almost a cleft chin. Every one of these features can be clearly seen in the photograph.

'As far as I can judge, despite the fur cap, the ears might also be Crabbie's. And in general build, the resemblance again is more than striking.

'I wouldn't have believed that Crabbie would consent

to work for the Russians but Mr. Hutton's book has convinced me. I have read and re-read the tape-recorded interrogation. The answers given by the man alleged to be Crabbie sound exactly the sort of things I should expect from him.

'As far as I can judge, the evidence appears to be irrefutable. I thought I had put the story of our life behind me, but now it has come to life again.'

* * *

The reaction by the British Press to *Frogman Extraordinary* was far from universally favourable. The *Sunday Pictorial's* front-page headlines on 29 May 1960 were:

'Crabb Storm. Is the frogman alive? Writer says: "He is now serving in Russian Fleet." Navy says: "This theory is ruddy nonsense".'

The report beneath stated: 'An amazing story which claims to reveal what really happened to frogman Lionel "Buster" Crabb, G.M., who vanished on an underwater spying mission four years ago, has roused anger in Whitehall.

'At the Admiralty a high-ranking spokesman declared: "Our experts have studied Mr. Hutton's exciting book and they have arrived at the conclusion that the *dossier* is a complete fabrication. The book is also discreditable to the Royal Navy and to the memory of a very brave officer and gentleman."

'The *Sunday Pictorial* has studied the story set out in the so-called secret Russian documents. And we are not satisfied that the documents on which the book is based are genuine.'

Jack Thomas, *Empire News* feature writer, who had known Commander Crabb and who had been one of the first to see both the *dossier* and the Russian photograph, made his own thorough investigation. Having read the *Sunday Pictorial* story containing the Admiralty statement he spoke to his own contacts there and on 29 May 1960, wrote in his newspaper:

'At the Admiralty last night it was stated: "No official statement on this matter has been given out by the Admiralty and neither has there been any official comment. So far we have been unable to find anyone here who has even read the book. We are making no comment".'

Jack Thomas concluded his article:

'Is the alleged *dossier* a fake? I don't believe it.

'I met Crabb several times, and to me his behaviour as described by his captors' reports, and his answers as recorded by the tapes, bear the ring of truth. *If they were faked, the man who faked them was a near-genius.*

'Listen——

'Col. Myaskov (interrogation judge): You do not regret what you did?

'Crabb: I regret I was caught.

'Myaskov: Would you consider buying your life by giving us information about Allied Naval secrets?

'Crabb: I would not. I am not a traitor to my country.

'Myaskov (raised voice): You damned silly English fool! Don't you realise that once we have pressed all your knowledge out of you we can do away with you?

'(Four and a quarter seconds' silence.)

'Crabb: Why haven't you done it already?

'Myaskov: Because we did not want to destroy your brain. We may have further use for you. But if you continue your damned silly nonsense, we will do what I said. I ask you for the last time. Will you do as I ask or not?

'Crabb: I will not. Drug me if you wish. Shoot me if you wish. But you can't persuade me to betray my country.

'That, as anyone who knew him would agree, is typical of "Buster" Crabb.

'Shall we ever get the truth about Crabb? The Russians are confident we shall not.

'I quote the final report from Capt. A. B. Nizhansky, chief security officer, operational command, Baltiysk :

'Called Lieut. Lev Lvovich Korablov (Crabb) to my office and informed him that he was "buried" on 5 July in Portsmouth. Let him read the English newspaper.

'From the way he behaved I am satisfied he is glad that his past is dead and that from now on only Lieut. Lev Lvovich Korablov lives.'

Jack Thomas is well-known for his writings on espionage and on the French and Danish Resistance movements. His book *No Banners** has become a standard work of reference in this field. It is unlikely that he would be deceived by an alleged *dossier* of a prolonged interrogation that was 'a complete fabrication'.

On the same Sunday, *The People* published the following :

'*Crabb.* Here is the latest tale about the missing frogman. *We say it's a phoney.*

'You can hardly blame Mrs. Margaret Crabb for having been persuaded last week to identify her husband, the missing frogman Lionel Crabb in a snapshot of some Russians in naval uniform. She has never believed that her husband died when he dived near Portsmouth Harbour four years ago.

'I have seen the snap. The figures in it are tiny.† No identification from it can be convincing. As "evidence" it is useless.

'As for the other evidence of Lionel Crabb's survival that Mr. Hutton has produced, I believe, quite frankly, that it is sheer poppycock.

'Having read the *dossier* and questioned Mr. Hutton about it, I have come to two conclusions. *One,* that Mr. Hutton firmly believes in its genuineness. *Two,* that it

* W. H. Allen & Company, London.

† The face on the enlarged photograph shown to the reporter of *The People* measured 1½ inches across.

is, nevertheless, a collection of incredible yarns, fit only for a cheap spy thriller.'

But the largest Sunday newspaper in Britain, the *News of the World*, carried on 29 May the front-page headline: 'Frogman Crabb *Is* Alive. Amethyst hero convinced.' Their reporter wrote:

'Commander J. S. Kerans, D.S.O., hero of the Amethyst's escape from the Yangtse and now Conservative M.P. for Hartlepools, told me last night: "*I am convinced Commander Lionel Crabb is alive—and in Russian hands. The Government must reopen this case.*"

'Commander Kerans has tabled a question in the House of Commons asking Mr. Selwyn Lloyd, the Foreign Secretary, to start new inquiries into the fantastic case of the British frogman who vanished when diving near a Russian cruiser in Portsmouth Harbour four years ago.

'His move to reopen the case follows a startling assertion by a former Czech journalist that Crabb is alive in Russia and working for the Red Navy under the name of Lev Lvovich Korablov.

'In a book to be published tomorrow the ex-Czech, Mr. Bernard Hutton, claims that a body washed up more than a year after Crabb vanished was "planted" from a submarine by the Soviet Secret Service.

' "I have examined the latest evidence very carefully," Commander Kerans told me. "I have discussed it with many naval friends with experience of espionage work. I do not believe this is a fabrication. I never believed the body buried at Portsmouth was Crabb.

' "Admiralty officials are said to have disclaimed the story that Crabb is alive. Of course they would do so. But to me the whole story has the ring of truth.

' "It is just the sort of thing the Russians do. I have studied the case very carefully and all the documents quoted by Mr. Hutton sound authentic. I believe the time has now come for an official statement on the whole affair."

'Crabb's former wife told me: "This book rings true".'

* * *

Commander J. S. Kerans, D.S.O., M.P., sent a copy of my book *Frogman Extraordinary* together with his own additional evidence on Commander Crabb to the Foreign Office and also demanded fresh inquiries into the case of the missing frogman in a question to Mr. Selwyn Lloyd, the Foreign Secretary. But he again met with failure.

The controversy continued.

A high-ranking officer of *H.M.S. Vernon* at Portsmouth, shown the photograph of Crabb-Korablov in Red Navy uniform, declared: 'This is *not* Commander Crabb. I knew him very well. He was a fine little man.' He insisted, however, that his name and rank must not be disclosed.

Mr. Edward Richman, proprietor of the *Sallyport Hotel* at Portsmouth, where Crabb had stayed before his last dive, said without hesitation when shown the snapshot of the group of Russian seamen: 'The one on the left is Commander Crabb. That's him, I'm sure. It is his face all right. I'm quite certain it is Commander Crabb.'

Miss Lily Cripps, the chambermaid to whom Crabb was a hero of heroes, at once picked him out from the group and said: 'Commander Crabb. He is wearing a strange hat—but it is him.' Asked if she were absolutely sure, she replied: 'Well, yes. Yes, it's him, quite definitely. I'm sure. I hope he's all right. He's a very nice man.'

Four naval officers who would not permit their names and ranks to be made known also without hesitation identified the man on the left of the Russian snapshot as Crabb.

But perhaps the most valuable identification of the frogman as the subject of the photograph and of the Soviet *dossier* was made by Chief Petty Officer Arthur

Ralph Thorpe, B.E.M., Commander Crabb's chief diving instructor and a very close personal friend since December, 1942—the man who had said: 'I've been with Commander Crabb on hundreds of missions'; the man who had met Crabb frequently after the end of the second world war.

In the course of a tape-recorded interview at his home in Portsmouth he discussed in detail Commander Crabb and the circumstances of his disappearance. Here are excerpts from that recorded interview:

Hutton: 'I'd like you to look at this picture* and tell me whether you know anyone on it.'

Thorpe (after several minutes): 'Well . . . looking at the different features, the chin, the nose and the eyebrows, I am almost certain that the man on the left is Commander Crabb. His nose and eyebrows have always been very familiar to me. I've seen him so often that way, really. When I've been studying the man when he's been in his diving kit with just his mouthpiece in, I've seen his nose and eyes more than I have anything else . . . I am quite certain it is Commander Crabb.'

Hutton: 'You are?'

Thorpe: 'Yes, I am. Yes.'

Hutton: 'You've read my book, *Frogman Extraordinary*. Would you tell me what you think about the Russian secret *dossier*?'

Thorpe: 'Well, I think that on reading your book I am convinced that Commander Crabb is still alive.'

Hutton: 'According to the Russian secret *dossier* Commander Crabb joined the Red Navy as a diver.'

Thorpe: 'Well, it's the job he likes . . . something he's interested in.'

Hutton: 'Knowing Commander Crabb very closely, did it surprise you, Mr. Thorpe, that he does his job excellently—or would you have expected him to sabo-

* The snapshot showing four men in naval uniform aboard a Soviet vessel.

tage it?'

Thorpe: 'Well, there aren't really any ways or means of sabotaging the job he's on, unless he's going to endanger somebody's life. He wouldn't do that.'

Hutton: 'Do you mean to say that even by having been forced to join the Red Navy he considers it his duty to assist anyone who is in trouble?'

Thorpe: 'Oh, yes, definitely.'

Hutton: 'Are there any comments you can make on the Russian secret *dossier*, Mr. Thorpe?'

Thorpe: 'Yes. There are many paragraphs in the book which contain typical ways of speech of Commander Crabb.'

Hutton: 'Typical?'

Thorpe: 'Yes. There are quite a few ways of speech which are most typical of Commander Crabb.'

Hutton: 'Would it be possible for an ingenious forger or an imaginative writer to have invented the *dossier*?'

Thorpe: 'No. No one could have invented it. No.'

Hutton: 'I have asked you this question because some people suggested that the whole thing might be a fake or hoax.'

Thorpe: 'I don't consider any of it a fake, because I've read it thoroughly and there are several remarks that Commander Crabb has made which are typical for him, as I said before. I know Commander Crabb well. I know how he behaves and how he speaks, and this is why I know that the book is genuine . . . that Commander Crabb is in Russia.'

After interviewing Mr. Thorpe I contacted the Chief Constable of Chichester, the Chichester coroner and the pathologist. I wanted to discuss the *dossier* with them. But they ignored my letters.

* * *

The arguments went on.

Some of the frogman's friends still insisted that 'Crabbie is decently buried and his grave is tended weekly. Why can't he be left to rest in peace?' But Patricia Rose and others close to Crabb stuck to their opinion that he was alive in Russia, willingly or unwillingly.

Mrs. Margaret Elaine Crabb told Danish newspaper reporters: 'The Russian *dossier* published in the book *Frogman Extraordinary* contains a number of outbursts and remarks which are most typical of my former husband and which no one could have been ingenious enough to invent.' Commander Kerans told the same journalists: 'I am convinced that Commander Crabb *is* alive in Russia. Doubtless he has suffered the usual brainwashing process. I have examined the evidence Mr. Hutton has published in his book and I am satisfied that all the documents are authentic.

'The headless body was a deliberate plant by the Russians—the same kind of trick we used ourselves during the war. Three days before the body was washed up at Chichester Harbour, Russian submarines were passing through the English Channel on their way to Egypt.'

The commander drew a chart of the currents in and around Portsmouth Harbour and said: 'The theory that Crabb's body could have been entangled somewhere and after some fourteen months freed by some strong current is unacceptable because, had this been the case, the current would not have taken the body to the spot where it was found.' And after talking about his own experience in Red China, gained while he was serving at the British Embassy in Peking immediately before the Amethyst episode, he continued: 'From my own experience of these matters and from my investigation, I believe Mr. Hutton's book is authentic.'

In most other countries of the Western world the publication of the *dossier* on Commander Crabb caused no controversy. It was accepted as authentic—not merely because Communist tactics are better known and understood outside Britain but because it was revealed that certain government departments in some of those countries had received from their own sources the contents of the *dossier* and knew that it was genuine.

In Great Britain, the full identification of the now famous Russian snapshot and the acceptance of the *dossier* by those who had been very close to Crabb made no difference to the official attitude. Nothing, apparently, would induce the authorities to reopen the Commander Crabb case.

Some people may still insist that *dossier* and photograph are clever forgeries. They are entitled to their opinion. Yet are not the men and women who knew Crabb intimately, and the unbiased investigators who conducted their own independent inquiries, checking and re-checking the evidence most thoroughly, the only judges really qualified to pronounce an expert verdict?

12

Third Officer Smith

The now accepted snapshot of the sailors on the Red Navy ship was published on the front page of the *Empire News*. Shortly after the appearance of *Frogman Extraordinary* the news editor of the *Daily Mail* contacted me and told me that a Mr. W. Smith, a haulage contractor in Stoke-on-Trent, had recognised his missing son (who had been reported dead by the Admiralty) among the four men in the picture. The news editor was anxious to enlist my help in trying to find out more about this other British naval officer who had disappeared mysteriously nine days before Commander Crabb was captured by the Russians.

I arranged to meet Mr. Smith at his home in Stoke and there was presented with all the available evidence about his son's disappearance.

Here are the facts, as he gave them to me:

Acting Third Officer William Russell Smith vanished without a trace from the Royal Fleet Auxiliary tanker *Wave Commander* on Tuesday, 10 April 1956. Next morning an officer of the Fleet Air Arm called to inform William Russell Smith's father that his son had been lost overboard and had not been found. The same day the Admiralty wrote:

'In confirmation of the verbal message given to you by our representative at Newcastle-under-Lyne, I regret to have to inform you that the Master of R.F.A. *Wave Commander* has reported that your son William Russell, acting third officer, disappeared overboard at about 7 p.m. yesterday in the English Channel whilst the vessel was en route to Port Said. Every endeavour was made by the *Wave Commander*, a Fishery Protection vessel, an Air Rescue launch, a lifeboat from Eastbourne and aircraft, including an R.A.F. Shackleton and a helicopter, to search the sea, but after 13 hours it is regretted that the search had to be given up as there was little likelihood of his surviving in the sea for so long and he unfortunately must be presumed lost at sea. There is just a remote possibility that he may have been picked up by a small vessel without wireless communication. However, should any further details come in we shall let you know at once . . .'

The naval investigation was held on Thursday, 12 April 1956, at Portsmouth. After having informed the parents about the extensive search which had been carried out in the area, the officer in charge repeated that there was a remote chance that the missing officer might have been picked up by a foreign boat not carrying a radio transmitter, but he added: 'That's only remote. I can't say that's authentic.'

Mr. Smith, senior, was assured again that he would be notified at once if there were any further development.

From that day onwards Mr. Smith and his wife watched the postman, hoping against hope that a telegram or letter would arrive from the Admiralty, informing them that their son had been found. They searched the newspapers and listened to the radio: but there was no further report. The only chance left was that Third Officer Smith might have been picked up by some ocean-going tramp and put ashore in a foreign port. Illness or loss of memory might explain his silence.

There was still the official inquiry to be held when the *Wave Commander* returned to Britain. This might shed more light on the mystery.

The *Wave Commander's* master, Captain Hubert George Carkeet, stated in his official report: 'At 19.50 hours, 10 April 1956, Seacunny* Ahmed Ali reported

* Lascar steersman.

to the bridge he could not find Apprentice Smith, Acting Third Officer.

'I was on the bridge with Chief Officer William Rennie and we thought Smith was with another officer, so the Chief Officer told the Seacunny to look in the Engineers' cabins. This he did and reported back at 20.00 hours that he could not find him anywhere. I took over the bridge and the Chief Officer organised a general search throughout the ship by all Ratings, Officers joining in.

'At about 20.15 hours, ship's speed was reduced to dead slow ahead and at 20.30 hours when the Chief Officer reported the search unsuccessful, the ship was turned round on reverse course. All available hands were stationed on look-out duties assisted with the aid of two 10-inch projectors.

'The Chief Officer remembered having seen Smith about 19.00 hours on the poop, wearing white shoes. They had been on the bridge together from 18.00 to 18.30 hours and had discussed Smith's sitting for the Certificate at the end of the voyage. They later dined together.*

'The Second Officer, John Edwards, was with Smith before dinner and remembered being asked some questions about distance from the land and future courses. He was then wearing black uniform shoes.

'The steward, Caetan Correa, saw Smith on the poop about 19.00 hours and wished him "good night," in answer to which Smith replied "good morning".

'Seacunny Abbas and Saloon Topass Joaquim Lopes were probably the last people to see Smith around 19.000 hours. He was then looking at the log on the poop.

'Apart from the disappointment of not getting home to see his parents, Smith seemed normal.

* At the inquest at Falmouth the First Officer and the Captain told Mr. Smith, senior, that they all dined at 18.15 hours (6.15 p.m.).

'At 20.47 hours an appropriate signal was promulgated to all ships and Niton Radio was asked to alert Coast Guard Stations. At 21.02 hours Niton was asked to alert the Royal Sovereign Lightship in the search.

'At 22.33 hours C-in-C Nore informed *Wave Commander* by signal that *H.M.S. Squirrel* and the Eastbourne lifeboat would join the search. Royal Air Force Rescue Launch and Shackleton aircraft assisting in the search at 02.30 hours 11 April 1956.

'At 03.30 hours after conferring with R.A.F. Launch and the Eastbourne lifeboat it was decided they should return to base, *Squirrel* and *Commander* continuing the search until daylight.

'At daylight a Royal Naval helicopter searched the vicinity with no result, and at 07.50 hours, the search was abandoned by *Squirrel* and *Commander,* but the helicopter continued the search.'

During the official inquiry, which was held at Falmouth in June 1956, and at which evidence in accordance with the report of the master of the *Wave Commander* was given, the coroner said he supposed that Acting Third Officer William Russell Smith had been lost overboard and must be presumed killed at sea. He would give a certificate to that effect. In the event of a body being found, Mr. Smith would be notified immediately so as to be able to identify it. The inquiry could be reopened, provided anything turned up.

Mr. Smith was given the opportunity of speaking with the *Wave Commander's* master, the officers and crew, and was given the following information :

At the time Acting Third Officer Smith was reported missing, the *Wave Commander* had been passing the *Royal Sovereign Lightship* off Eastbourne.

The master said there was a fast tide going to the English coast and he (Smith) would probably get ashore quickly if he missed the *Royal Sovereign.* The sea was calm and at the time Smith was reported missing the *Wave Commander* was about four miles from Beachy

Head. *No body which would fit William Russell Smith's description was ever found.*

As time went by and no new information came to light Mrs. Smith, who, in her husband's words, had been 'as fit as a fiddle,' fell ill and died. Her husband, her son Donald, relatives and friends believed that Russell's disappearance had helped to put her in her grave.

Then, on Sunday, 29 May 1960, Mr. Smith's heart was filled with fresh hope.

Donald Smith saw on the front page of the *Empire News* the picture of the group of Russian sailors. He showed it to his father and, pointing to one of the men with Commander Crabb said : 'It's like our Russell, Dad.'

Mr. Smith recognised his elder son at once. 'That's my lad!' he exclaimed. 'That's your brother! That's my lad or my name's not Smith!'

He was so sure that the man in Russian naval uniform was his missing son that he showed the picture to his sister. She also was convinced that the man was her missing nephew. There was a similar and immediate reaction from Mr. Smith's sister-in-law and from several friends who had known Russell since boyhood.

After four long years of nerve-wracking anxiety Mr. Smith was determined to get in touch with his son at last. He wrote to the Admiralty, to the Foreign Office and to the Soviet Ambassador in London, asking them to find out the young man's whereabouts.

The first reply came from the Embassy of the U.S.S.R. Dated 25 June 1960, the letter stated : 'I have been asked by the Ambassador to acknowledge the receipt of your letter to him of 10th instant, and to inform you that the Embassy has no information whatsoever on the question you raised in this letter.'

Though the Soviet Ambassador in London did not refute the possibility that William Russell Smith was indeed the man in the photograph, Mr. Smith was disappointed that no offer was made to try to check with Moscow.

The next communication came from the Admiralty. Dated June 30, 1960, it stated: 'In reply to your letter dated 10th June, 1960, addressed to the First Lord, I am commanded by my Lords Commissioners of the Admiralty to inform you that the fact pertaining to your son's disappearance from R.F.A. *Wave Commander* has been reviewed. They consider that there can be no possibility of his having joined a Soviet ship.

'2. My Lords deeply regret that they can do no more to allay your grave anxiety.'

Although this letter excluded all hope that William Russell Smith was alive, Mr. Smith and his family were still convinced that it was their Russell who appeared in the *Empire News* picture. They hoped that the awaited reply from the Foreign Office or news from behind the Iron Curtain might be more encouraging.

The Foreign Office asked Mr. Smith for the photograph of the man claimed to be his son so that appropriate inquiries could be instigated. Having obtained the requested print, the Foreign Office took every step to clear up the mystery. In a letter dated 1 November 1960, Mr. Smith was told: 'I am directed by the Earl of Home to refer to your letter of the 19th August about the photograph you had seen in the Sunday Empire News.

'Her Majesty's Embassy in Moscow were asked to take up with the Soviet Ministry of Foreign Affairs the suggestion that one of the persons in the photograph may possibly be your son.

'The Soviet Ministry of Foreign Affairs have now replied that the tracing of missing persons in the Soviet Union is a matter for the Red Cross and similar organisations and should be made through the National Red Cross Society concerned.

'This Department have therefore communicated with the British Red Cross Society of 14-15 Grosvenor Crescent, London, S.W.1, and they have agreed to take up the case. I am to suggest that you should also write to this address, marking your letter "for the attention of Miss Wallbridge."

'Your copy of the photograph is being returned to you as you requested.'

Meanwhile, on 29 October, 1960, an extensive report from Iron Curtain informants whom I had asked to look into the matter reached me in England. Enclosed were two photographs of men in Soviet uniforms. I told Mr. Smith the same day and, on meeting him on 3 November, I acquainted him with the contents of the document.

This report contained a description of the man in the photograph—date and place of birth, height, colour of eyes and hair, and even such particulars as a scar on the right cheek, an appendix scar, and rather large hands—which Russell's father acknowledged to be absolutely correct. The report confirmed that Smith was in Soviet Russia, that he was serving in the Red Navy under the identity of Second-Lieutenant Rodion Vladimirovich Smirnov, and explained how the transformation had come about.

In 1954, William Russell Smith—not then eighteen years old—was picked for recruitment by 'Hendrik', the Soviet resident spy operator in Port Said. 'Hendrik' hoped that the youth would be able to supply him with useful information about the Royal Navy. He introduced Smith to his associate 'Vera', whom he claimed to be his niece. On her superior's instructions 'Vera' succeeded in making the young, inexperienced Smith believe that she was in love with him.

One evening she took the youth to a club, where she managed to drug him. When he came around, he discovered that his wallet had been stolen. 'Hendrik' arrived and persuaded Smith to accept a loan of £50. For this he obtained a receipt which was later doctored to 'confirm' that the amount of £50 had been paid 'on account of payment for information about British Naval secrets'.

From that moment William Russell Smith was filed in the Central Index at Moscow Secret Service headquarters as *Nash,* meaning 'He is ours.' His first refusal to betray his country was soon broken down by blackmail and he was ordered to pass information to a Soviet agent in England.

During the trip back to England Smith determined to defy 'Hendrik's' threats but he made the mistake of telling the Russian agent in England that he intended expose the plot and confess everything to his commanding officer. A coded radio message was sent to Moscow and the agent was ordered to get the young man into the hands of the 'transport department'. It was vital that the Soviet espionage network in Port Said should not be revealed to the British authorities.

When he met Smith again the agent told him that he had bad news. He had received information that 'Hendrik' had been arrested. 'Vera' was safely on her way to the Soviet Union but 'Hendrik' was trying to save his own skin by pretending that he had been acting only under the girl's orders. Unluckily, the British authorities had found in 'Hendrik's' office documents which implicated a number of other people, all of whom were being hunted down. The file on Smith had been found; his commanding officer knew all about it and had been told to return him to Port Said so that he could be arrested and charged.

Smith, already suffering considerably from remorse and guilt, never doubted the tale. His first reaction was that it would be best to give himself up, make a full statement on how he had got caught in the spy network, and take his punishment. The astute agent soon convinced him that this would not help. As far as the Royal Navy was concerned he was already condemned. To Smith's other suggestion—that he should jump ship and try to get away from England to start a new life elsewhere—the agent replied that the thought was childish. As soon as he was posted as a deserter Interpol would be put on his track. It would not take long to find him.

'Then what, in God's name, am I to do?' the youngster asked, panic-stricken.

The agent's tone changed. He became positively fatherly. The Soviet, he said, would take care of everything. Smith should behave quite normally. He should sail in his ship when it left the Isle of Grain. At a time which would be arranged later he should dive overboard and swim to a point where a boat would be waiting. He would be taken across the English Channel and his further transport to the U.S.S.R. would be arranged. Once safely in Russia he could join the Red Navy, in which promotion was easy and swift. Soon he would be reunited with his friend, 'Vera'.

After thinking things over carefully, Smith decided no other way was open to him. He would do as he was told. At least this solution would save him and his family from disgrace.

When the *Wave Commander* sailed from the Isle of Grain on 10 April, 1956, he prepared for his planned disappearance by telling members of the crew that he felt sick. At the proper time and place he dived overboard unobserved. Since all his possessions were left in his cabin there was no reason to think that his disappearance would be reported as anything but an unfortunate accident.

He was a strong swimmer and, with a strong tide to assist him, he had no difficulty in making his way to the waiting vessel. After being landed on the other side of the Channel he was transported via East Berlin to the U.S.S.R., and the danger to the 'Hendrik' circuit in Port Said was removed.

On 8 October, 1956, Smith was given identification papers in the name of Rodion Vladimirovich Smirnov and sent for elementary training to the First Ukrainian

Command. On 22 November he was posted to Sevastopol and enlisted into the Black Sea Fleet as an ordinary seaman. His service was so satisfactory that by 25 March, 1957, he had already been promoted several times. On 18 December, 1958, he was transferred to the Far Eastern Naval Command, where he remained for ten months. There he met Crabb, alias First-Lieutenant Lev Lvovich Korablov, and served with him for almost eight weeks. On 31 October, 1959, he was transferred once more to the Black Sea Fleet and received his promotion to the rank of Second-Lieutenant on 7 November, 1959.

* * *

That was the report which I showed to his father in his home at Stoke-on-Trent.

'He was a fool!' Mr. Smith exclaimed when he had read it. 'Russell could easily have been blackmailed if the blackmailer struck at the right time and struck the right tune. Yes . . . he would fall for this kind of threat. He behaved very foolishly.'

He looked for a long time at the photographs which showed his missing son among other men in Russian uniforms. 'It's our Russell in both pictures,' he said, pointing to one of the figures. 'Yes, it's our Russell.'

He picked up the report again and re-read several passages.

'I knew it was our Russell when I first saw the picture in the *Empire News,* and so did our Donald and all the other relatives and friends,' he said with great emphasis. 'I could never figure out how Russell could have been killed at sea. The master of the *Wave Commander,* Captain Carkeet, told me that there was a fast tide going to the English coast and Russell would probably have got ashore quickly if he'd missed the Royal Sovereign lightship. Russell was a very strong swimmer and could have done the short distance easily, especially if helped by the tide. And don't forget—no body was ever found which could fit Russell's description.'

He continued to study the pages. Then he said:

'Yes, I'm convinced that every word in this Russian report is true. It makes sense. Everything ties up. I can see Russell behaving like this clearly from it . . .

'You know, he phoned me from the Isle of Grain when the *Wave Commander* returned from Port Said and told me he couldn't get leave to come home to see us. He was very eager to see us and was very despondent that he wasn't allowed to leave, even for a flying visit. I only wish I'd gone to see him at the Isle of Grain. Perhaps he'd have confided in me and the whole affair might have been settled.'

I doubted that, knowing how thoroughly a trained agent can instill fear; but I said nothing.

'Let's hope we may get some more news—a direct message from him, perhaps,' Mr. Smith sighed. 'Let's pray we may see him again.'

As it happened, I was able later to get a little more information from my contacts about the career of Acting Third Officer William Russell Smith, alias Second-Lieutenant Rodion Vladimirovich Smirnov and his father was happy to know that his son was alive and well; but his hope to see Russell again was unfulfilled. Mr. Smith died suddenly in December, 1964.

13

Icebreaker In Difficulties

Then months after the publication of *Frogman Extraordinary* further important information about Commander Crabb reached England. The report from the U.S.S.R., which was accompanied by another photograph of Crabb-Koroblov in uniform aboard a Red Navy vessel, gave full details of his activities in Russia after the last date recorded in the original secret *dossier.*

It stated:

'On 11 February, 1960, First-Lieutenant Lev Lvovich Korablov was summoned before the commander of the Far Eastern Naval Command, Vladivostok, and informed that on orders from Khimky Naval Intelligence Station he was to be given special leave and flown to Moscow. Due to unfavourable weather conditions First-Lieutenant Korablov had to remain the following two days in Vladivostok, however. He did not return to his Command but stayed at Vladivostok headquarters.

'On 14 February, 1960, First-Lieutenant Korablov was flown to Moscow, where he was met at the airport and driven to Khimky Naval Intelligence Station. From there he was taken to Investigation Judge Lieutenant-Colonel Boris Fyodorovich Yevdokimov. No exact record of the full conversation is available but it has been established that the Investigation Judge informed First-Lieutenant Korablov about world-wide Press statements revealing that Commander Crabb is alive in the Soviet Union. Korablov's sudden summoning to Khimky was explained as being a security measure to prevent any foreign Press or government informer recognising him and passing this information to his chiefs.

'First-Lieutenant Korablov, who, it is stated, insisted that he would always deny any suggestion that he is indeed Lionel Crabb, was put on "indefinite leave" and sent to the Recreation Home in Yalta. The summing-up of the Political Officer of the establishment complains that "First-Lieutenant Korablov behaved quite unsociably and stubbornly kept himself to himself."

'On 19 March, 1960, First-Lieutenant Korablov was ordered to report to Black Sea Fleet Command at Odessa. There he arrived on Sunday, 20 March 1960. The Commanding Officer told him that for the time being he would be in charge of the Underwater Diving School, where he was to hold a special diving instructor's course on difficult underwater manoeuvres. Korablov commenced with the course forthwith but his weekly reports to the Commander-in-Chief stated that all his pupils were well-trained frogmen, able to carry out any given task, and that there was not much he could teach them. Despite this honest assessment of the abilities and qualifications of these men, Korablov was nevertheless ordered to continue and carry out his given orders.

'On 31 May 1960, First-Lieutenant Korablov was summoned by the Commander-in-Chief and informed that on orders from Khimky Naval Intelligence Station he was to fly to Moscow. He left Odessa by air at 10.55 hours the same morning.

'On his arrival in Moscow First-Lieutenant Korablov was met at the airport and driven to the Department of Information of the State Security Committee, where he was acquainted with the fact that full details about him—from the day of his disappearance in Portsmouth on Thursday, 19 April 1956, to his posting to Vladivostok the previous year—had been published in London. He was also shown British newspapers and his picture in the Red Navy which had appeared on the front page of the *Empire News* and which had been identified by his former wife.

'Korablov carefully read the newspapers but did not comment. When asked what he had to say, he replied that only some newspapers accepted the details for a fact while others expressed their disbelief. He agreed, however, that his photograph was clearly recognisable but added that although this was the case, some London newspapers had nevertheless not accepted it as conclusive proof.

'Intensive measures were taken to find those who had been responsible for this serious leak and State Security officers left for the German Democratic Republic (East Germany) to investigate the matter on the spot again. As far as First-Lieutenant Lev Lvovich Korablov was concerned, security measures were tightened considerably and he was given strict orders to make sure that under no circumstances must he be photographed.

'On 3 June 1960, First-Lieutenant Korablov was flown back to Black Sea Fleet Command, Odessa, and was from then on attached to Operational Command. His duties were those of Commander of Diving Operations.

'On 26 October 1960, First-Lieutenant Korablov was transferred to the Northern Fleet, Baltic Sea Fleet Command. His name was contained in the Awards List issued in connection with the 43rd Anniversary of the Great October Revolution and he was promoted to the rank of Commander. At a ceremony in Leningrad on 17 November 1960, First Deputy Commander-in-Chief of the Soviet Navy, Admiral Arseniy Golovko, presented him with a Government Order and medal for his "bravery and selfless service which not only saved lives but also a very valuable vessel."

'Commander Lev Lvovich Korablov continues serving with the Baltic Sea Fleet Command. On 21 November 1960, he was temporarily posted to Estonian Operational Naval Command to supervise special underwater operations, but when his task is completed he is to return to his duties as Commander of Diving Operations at Baltic Sea Fleet Command.

'During his four and a half years of his stay in the U.S.S.R. Commander Lev Lvovich Korablov has not become a member of the Communist Party of the U.S.S.R., but everyone of his Chief Security and Commanding Officers, as well as his fellow-officers and men under his command, considers him a genuine comrade whose one and only aim is his work and who is always ready to help anyone who may need his assistance. It is a well-noted observation that he undergoes frequent melancholic spells, during which he hardly talks and smokes and drinks very much more than usual, which means that his tobacco and vodka consumption is very considerable because he always smokes heavily and drinks a fair amount. His friends assume that something from the past must be weighing heavily on his mind, but every attempt to gain his confidence is futile and he just keeps himself to himself. Frequent medical reports state that he is remarkably fit for a man of almost 50.'

This report, smuggled out of Russia, was in some parts corroborated by officers and men of the Red Navy who had served with Korablov. They spoke admiringly of the 1 metre 67 centimetre (5ft. 6in.) tall officer 'who speaks with a somehow strange pronunciation,' and were unanimous in their verdict that 'despite being somewhat aged for special diving and frogman duties, he nevertheless is as capable under water as any young and first-class frogman'.

During a friendly visit of Red Navy vessels to Scandinavia, a second-lieutenant who had served with Korablov in the Far Eastern Naval Command, Vladivostok, told why Crabb-Korablov had been promoted to the rank of Commander and presented with a Government Order and medal.

'In January 1960, while we were carrying out a routine naval exercise, we picked up an SOS from an icebreaker in difficulties,' he said. 'The icebreaker's skipper reported serious damage to the rudder and made it clear that immediate help was imperative. Otherwise he would be unable to leave the danger area and would risk being crushed by rapidly forming ice. We proceeded immediately to the icebreaker but were unable to get near enough to take her in tow. We had no icebreaker escort and our ship was not strong enough to risk getting through the already heavy ice layers.

'Our Commander-in-Chief sent a reconnaisance party over the ice to find out whether we could render the crippled vessel any assistance and help her to safety, but it was ascertained that the only possibility was to go down and repair the damage under water. This seemed impossible because the sea under the frozen layers was ice-cold. Everybody feared that if a frogman or diver attempted to carry out the repair on the rudder, he might freeze to death. The only solution our Commander-in-Chief could see was to take the icebreaker's crew and most valuable equipment aboard our ship.

'Comrade Lev Lvovich Korablov opposed our Commander-in-Chief. He insisted on being permitted to survey the damage. He was stubborn and eventually succeeded in getting our Commander-in-Chief to change his mind.

'As he put on extra woollies in preparation for the suicidal dive he asked for a treble ration of vodka. Then he dived. After a number of dives into the icy water—and a corresponding number of treble vodkas—he managed to replace the broken bolt in the rudder and make the icebreaker manoeuverable again.

'Everybody cheered Lev Lvovich Korablov as the icebreaker broke away from the danger area—saved from being crushed by the ever-mounting ice formations. Korablov became a hero talked about throughout the Far Eastern Naval Command and other commands, and an account of his incredible deed was published in the Soviet newspapers. A month after the icebreaker episode Korablov was transferred but he will never be forgotten in Far Eastern Command.

'When Comrade Korablov was promoted to the rank of Commander last November, and when, at the great naval ceremony in Leningrad he was presented with the Government Order of Honour and the medal, most of us old colleagues of his were invited to the ceremony. We all were more than proud that *our* Lev Lvovich was officially honoured for his heroic act which had saved a valuable ship and many lives. But he was embarrassed when he was publicly proclaimed a hero and repeated over and over that he had done only what any other diver or frogman would have done in the circumstances.'

Chief Petty Officer Arthur Ralph Thorpe, B.E.M., probably knew 'Crabbie's' service behaviour and character better than any other man. When I told him the story of the icebreaker, he said: 'That's most typical of Commander Crabb. Remember I told you that he would always do everything in his power to assist anybody who needed help? That icebreaker job was a sort of challenge for him—just his cup of tea, so to speak. It reminds me of so many daredevil episodes he volunteered for during the war. Yes, it's most typical of the Commander. It firmly convinces me that your Red lieutenant didn't invent a word but spoke the strict truth.'

* * *

During 1961 reports from the U.S.S.R. and other Iron Curtain countries provided more and more irrefutable evidence that Crabb was alive and serving in the Red Navy. Each report originated from an impeccable, proven source—every one unconnected with any other—and confirmed the facts again and again.

One of the most important and convincing revelations came in July that year from no less a person than

Sir Percy Silitoe, K.B.E., C.B.E., former director of Britain's M.I.5.

He said: 'The Soviet secret *dossier* published in the book *Frogman Extraordinary,* is genuine. The exposure of the secret *dossier* in the book did not surprise anyone in Intelligence because, some considerable time earlier, the complete text of the *dossier* was obtained by Intelligence agents, and both texts are identical.'

Sir Percy revealed that he himself had not been surprised by the information given in the document. 'At the time of Commander Crabb's disappearance,' he said, 'it was known—because there are eyewitness accounts for this—that he did not perish but that he had been intercepted by Russian frogmen who were in the water around the Soviet cruiser. The fact that Commander Crabb did not perish in Portsmouth Harbour was also confirmed by coded Russian radio messages which were intercepted and decoded and which revealed that a prisoner was being flown to Russia. The rendezvous between a Russian helicopter and the Russian cruiser in the Skagerrak which was witnessed by a Danish destroyer captain and authenticated in Copenhagen, and trustworthy statements from Red Navy officers who confirmed that a man had been captured in Portsmouth, had been held prisoner in the cruiser and had been flown by helicopter to Soviet territory, amply corroborate and indeed confirm that Commander Crabb did not perish.

'The ultimate finding of the headless and handless body near Chichester proved the fact. The size of the body did not match exactly the size recorded in Commander Crabb's Royal Navy records, and a thorough study of the first and ultimate pathologist's reports establishes that there are too many discrepencies to allow anyone to say: "This *is* the body of Commander Crabb." It could well be that the day may come when the authorities could be forced to reopen the case, and if this happened much interesting information might come to light.'

Six weeks later, my colleague and friend, Johnny Hunt-Crowley, a well-known newspaper reporter, told me that he had interviewed a naval lieutenant whom he had known well for a considerable time and who had told him: 'Crabbie is not dead. I know exactly what happened to him.'

'What do you know?' Johnny asked.

'Well, together with' (he named another officer) 'I watched Crabbie dive into the water on 19 April 1956, and saw him swim towards the spot where the Russian cruiser berthed. Later on we saw Russian guards dive into the water and intercept Crabbie. After a brief struggle they disappeared under water and none of them reappeared. There's no doubt in anyone's mind that the Russians must have taken Crabbie through an underwater airlock aboard their cruiser.'

'Yes, but how do you know the frogman was Crabbie?' Johnny asked.

'He was a close friend of mine. I knew his figure and his frogman suit. We watched him through binoculars and had a perfect view.'

'Did you report it at the time?'

'Of course we did. We were asked to submit a written report, and that's what we did.'

'You're a fine pal,' Johnny said. 'Why didn't you tell me at the time? It would have made a sensational front-page story. We'd have scooped the world.'

'Don't be daft!' his friend retorted. 'We were pledged to secrecy. Hell! you know enough about the Official Secrets Act to know what would happen to a serving officer if he didn't keep his mouth shut. And for that matter, what I'm telling you now is confidential—not for your blasted newspaper.'

Johnny was still sceptical. 'You claim you know exactly what happened to Crabbie,' he said, 'but you've only told me that you and your mate watched him being captured. For all you know, he could have been drowned.'

'Not Crabbie! Besides, there's evidence that he was taken to Russia and that he's serving in the Red Navy.'

That was as much as Johnny could get out of him.

'Are you satisfied that your friend was telling you the truth, Johnny?' I asked him.

He bristled. 'You ought to know better than to ask such silly questions. Would I have told you, if I weren't?

There was still another confirmation that Crabb's last dive had been seen. George Rodney (Bob) French, another newspaperman, said in December 1961 that he had spoken to a naval officer,* and that this man had given him an identical version to that which Johnny had heard.

* * *

The Soviet authorities were angered and alarmed by the persistant security leakages but decided to ignore all statements made in the West—at least, for the time being. However, Moscow's hand was forced when Red Navy Lieutenant Vadim Andreyevich defected on 5 January 1962, and revealed that he had served with Commander Lev Lvovich Korablov in the Baltic Sea Fleet Command. He provided such precise details about the frogman that there could be no doubt that Crabb and Korablov were one and the same.

It was agreed at a high-level State Security conference in Moscow that the time for action had come. It was decided not to reissue any of the nine previous statements that Crabb was dead but to use subtler methods which the State Security directors hoped would 'bury' the frogman once and for all.

At the beginning of April 1962, an 'authentic report from Russia' was leaked through Berlin. The report purported to come from bona fide informants, reached the West through an uncheckable route, and stated:

* **The fellow-officer of the lieutenant whom Johnny knew.**

'Commander Lionel Kenneth Philip "Buster" Crabb, O.B.E., G.M., R.N.V.R., alias Commander Lev Lvovich Korablov, will never see his native country again. Not because the Soviet Government would prohibit him from leaving the U.S.S.R. after his term of service in the Red Navy expired; not because after ten years' stay in the U.S.S.R. he might perhaps no longer wish to return to England; but because the 167 centimetres-tall frogman-hero of World War II is now dead.

'Commander Crabb's death in early 1962 is surrounded by as much mystery as was his sudden disappearance on 19 April 1956, in Portsmouth Harbour. For, due to the fact that neither the Khimky Naval Intelligence Station nor the Red Navy nor the State Security Committee recorded the circumstances of Commander Lev Lvovich Korablov's death, but only registered that "he died on active service", it has been impossible to discover the full facts, despite every effort having been made to provide the last chapter of the Commander Crabb mystery.

'This is all that has been established:

'On 3 February 1962, news reached Crabb-Korablov's former comrades at the Far East Naval Command, Vladivostok, that "Comander Lev Lvovich Korablov died while carrying out a diving mission." Two days later, on 5 February 1962, his death was announced in a Red Navy Order of the Day, which also stated that the frogman had been buried with full naval honours.

'Crabb-Korablov's naval officer colleagues did their utmost to find out the circumstances of his death. The information they received from his commanding officer stated that "the frogman died because of lack of oxygen, caused by a faulty valve. He was lost while carrying out an underwater inspection of a destroyer."

'The frogman who claims to have accompanied Commander Lev Lvovich Korablov on his last diving mission, however, insisted: "Commander Lev Lvovich Korablov and I carried out an underwater survey of a submarine which had failed to surface. Commander Korablov was trapped in the submarine's tower and could not be rescued in time."

'The security officer of Crabb-Korablov's unit stated: "The frogman contracted pneumonia and died in the Naval Hospital in Kronstadt." The hospital records do not, however, contain the name of Commander Lev Lvovich Korablov—neither in the admission list nor among those who died there.

'Crabb-Korablov's frogman colleagues who served with him in the Red Navy were unable to give an exact account of how the frogman died. All they were able to say was that at the Red Navy Baltic Sea Fleet Command of the Northern Fleet, in Leningrad, the 167-centimetres-tall Korablov had lain in state in his frogman suit, surrounded by flowers, and that the commanding officer spoke highly about the comrade-hero whose tragic death was a great loss to the Red Navy. They also confirmed that their dead colleague was given a funeral with full naval honours.

' "The sudden death of our comrade, Commander Lev Lvovich Korablov, deprives us of perhaps the greatest frogman the world ever had," said Admiral Golovko, the First Deputy Commander-in-Chief of the Red Navy, at Korablov's funeral. "Everyone who knew him and worked with him will continue to treasure the dear comrade and friend. He will live in our memories. We all will follow in his footsteps. Our dead comrade who was so small in body but so great in his heart and work will never be forgotten."

'Independent investigations into the circumstances of how Commander Crabb-Korablov met his end did not meet with greater success, either. Consequently it is at present impossible to establish exactly how he died. It is rather doubtful whether exact details will ever be unearthed.'

This 'authentic report from Russia' did not ring true to me. I studied every word of it. The whole presentation was completely different from any other communication which I had received from Iron Curtain sources. The fact that it came from unknown and uncheckable informants made me suspect that it could be a Soviet State Security 'plant'. The only way to find out the truth was to check with my old and trusted informants in Europe. I acted immediately.

Six weeks later I received a message from one of my most impeccable contacts and it reached me by the customary route. It stated categorically that 'Crabb-Korablov has stayed at Sevastopol between 15 February and 10 March 1962, when he returned by air to Leningrad.' My doubts had been justified.

Months of frustrating waiting went by; months during which my informants remained silent. Past experience had taught me that illegal investigations in the U.S.S.R. often require many months and that sometimes even the most concentrated efforts proved fruitless and long-awaited information just could not be obtained. Then, in mid-September 1962, news arrived.

'The report about Korablov's death is a falsification,' my contact stated. 'No Red Navy Order of the Day of 5 February 1962, or any other day, contains an announcement about a frogman, naval officer or seaman having been buried with full naval honours. At the Red Navy Baltic Sea Fleet Command of the Northern Fleet in Leningrad no frogman or any other person has lain in state at any time. There is no record of the speech Admiral Golovko is alleged to have made but it has been established that he was not at the purported funeral because during the time in question he was not in the vicinity. Commander Korablov is alive, in good health and on continued service. At the time of his alleged funeral he was in Murmansk. From mid-February to 10 March 1962, he was in Sevastopol. He still serves with the Baltic Sea Fleet Command.'

In November 1962, another genuine message was received. It told me :

'Commander Lev Lvovich Korablov attended the festivities of the 45th Anniversary of the Great October Revolution in Kronstadt.'

Further authentic reports about Commander Korablov's movements and service continued to reach the West at irregular intervals throughout the rest of the year.

14

The Woman From Turkey

Patricia Rose also had received through the years a series of messages alleged to have come from her fiancé. Her doubts whether these mysterious communications were genuine had been fairly quickly dispelled. She had been given so many details about things which only she and Crabb knew that any suspicion of deceit could be ruled out.

While waiting to take part in a *Granada* television programme about the Crabb affair she met in the studio a woman who talked about her 'beloved Turkey'. After the programme had gone off the air, this woman told Pat that she had followed the Crabb mystery closely, adding that she understood well the feelings of someone who had been parted from a loved one.

She told Pat she was sure that everything would turn out well in the end. Then she said: 'You may know this already, since you meet so many people, but there's no harm in talking about it. When the photograph of Commander Crabb in Russian uniform was published in *Paris Match* Martin Solomon, who knew your fiancé very well, said that he was fully convinced that the picture was indeed that of his friend.'

'What photograph?' Pat asked. 'I don't know anything about it.'

'Well, it was said that the picture had been smuggled from Russia and that it had been taken in 1959 on board a Russian naval vessel—in Vladivostok, I seem to remember. *Paris Match* published a huge enlargement of the man who was supposed to be Commander Crabb and Martin was positive that it was he. I think we can take it for granted that he was right. Martin was surely able to judge whether it was his friend or not.'

'I must try to get hold of a copy,' Pat said.

'You should—so that you can judge for yourself. I suppose it *is* possible even for Martin to be mistaken.'

During the conversation it came to light that the woman was connected with a Middle East anti-Communist underground movement with unique contacts inside the U.S.S.R. and satellite countries. She promised to keep in touch with Pat and, if possible, to get further information for her.

Some time later Pat received a letter from the woman. She was in Rome, where, she said, she had met a man with important connections in the Communist world. He had told her: 'Though many people thought that Commander Crabb was dead, this is untrue. We know for a fact that he is alive and that he is a high-ranking officer in the Red Navy.'

The woman promised to find out more and try to put Pat in touch with the underground movement so that she could keep in direct contact and receive information with the least possible delay.

A few weeks went by; then Pat receive a telephone call from a man who referred to the letter and suggested that she should meet him in London.

She kept the appointment at the *Grosvenor Hotel*, Victoria, and he told her: 'I can assure you most definitely that Commander Crabb is alive and is serving with the Red Navy. The information I am to give you has been confirmed and reconfirmed and there is the fullest evidence that it is authentic. I am also to tell you that you will be seeing your fiancé again.'

He gave her precise personal details which proved to her conclusively that, directly or indirectly, he and his movement must be in touch with Crabb. Before leaving her, he said: 'You should get in touch with the people who work for the underground movement called *Blok*. They should be able to get quite a lot of additional information for you.'

Through her friend Captain Brockerbank Pat met a Pole who introduced her to a member of the underground movement at the Polish Club in London. This man told her that, with two Polish generals who were now living in the United States of America, Commander Crabb had been active for *Blok*. He promised to help her to find out as much as possible about her fiancé.

Six weeks later another member of *Blok* contacted her. 'There is definitely no need for you to worry about Commander Lionel Crabb,' he said. 'It is an established fact that he is alive. He now holds the rank of commander in the Red Navy and he is in charge of the training of special task divers and frogmen. He wants you to know that he is longing to be reunited with you and that one day your wedding will take place.'

He gave the girl personal details which none of her previous informants had known and which were so precise that they startled her. He repeated the facts about Crabb's capture and what had happened to him afterwards. His story tallied in every respect with what she had already been told by her other anonymous friends.

'I am most grateful to you for all that you are doing,' she said. 'But why can't we ever get a note from Crabbie in his own handwriting? It would be such irrefutable proof.'

'I understand your desire, but it can't be done,' he replied. 'It's too dangerous. You must realize that your fiancé is not in England, where anyone can write and send as many letters as he wishes. He is in Soviet Russia, where he is under constant surveillance by State Security . . . the secret police, you know. If he should make a wrong move and it were discovered by State Security, it would mean the end of the road for him. You must realize that all approaches to him have to be made with the greatest care and in utmost secrecy, so as not to endanger him in any way. No, you must be content with the information you receive indirectly. The main thing is that you know your fiancé is alive and well.'

'But he once sent a message in a bottle,' she protested. 'If he only knew how happy his mother was when she saw and recognised his handwriting, he——'

'That was a very foolish and dangerous thing,' the man cut in. 'Your fiancé can thank God that he was not caught and punished with the supreme penalty. That piece of bravado occurred before he realised fully how dangerous it is in Soviet Russia to try to smuggle out any secret messages. However, from the information we have, he now knows exactly how to behave for survival and, having accustomed himself to the secret-police type of life, he would not dare to repeat the mistake which could have cost him his life.'

* * *

The Crabb affair was revived during a friendly visit by Soviet warships to Egypt. Two seamen who had been in the cruiser *Ordzhonikidze* while the Soviet flotilla was anchored in Portsmouth Harbour in April 1956, bragged over their drinks about how an English frogman had been captured and taken aboard the flagship on the second morning of their stay in the British port. Fuddled by bad liquor, they did not notice that the man to whom they were speaking had a *Miniphon* wire-recorder with a microphone concealed beneath his tie; nor did they suspect that he was a newspaperman.

'Did any of you actually see this captured frogman?' he asked.

'No, we didn't see him. We were on duty on the other side of the deck,' one of the seamen replied. 'But we heard from comrades who saw him that he was a very small man with a protruding nose and very thick eyebrows.'

'He was taken immediately to the hospital quarters,' the other Russian chipped in. 'He was probably hurt.'

'So you didn't see him—and you don't know what happened to him?' the reporter said sceptically.

'Not really—but he must have been ill because when we were on the homeward voyage a helicopter took him and one of our officers off the cruiser. Someone said he was being flown to a hospital ashore.'

'How do you know he was an *English* frogman?'

'He must have been. We were in a British port,' the first seaman replied. 'Our frogmen wear completely different diving suits. It was certainly not one of our comrades.'

'Maybe he was a frogman from one of your destroyers,' the reporter suggested. 'If he was, you wouldn't know him.'

'I told you . . . our frogmen wear completely different rig,' the Russian insisted. 'He was an Englishman. That's clear.'

'If he was an Englishman, why wasn't he handed over to the British authorities?'

'I don't know. Maybe he'd come to ask for political asylum.'

'But you said he was probably hurt. Wouldn't it have been wiser to send him ashore to a British hospital?'

'If he had come to ask for political asylum he wouldn't have wanted to be sent back to a British hospital,' the Russian said. 'Maybe he had his reasons.'

'Couldn't this frogman have been on a spying mission on the underwater construction of your cruiser? Couldn't he have been intercepted and captured by your own frogmen?'

'If he had been a British spy he wouldn't have been taken to our sick-bay and flown later to a shore hospital. He would have been killed. Espionage is punishable by death in our country.'

'But you were in a British port,' the reporter pointed

out. 'If the frogman was a spy, he would have to be dealt with by the British authorities.'

The Russian laughed. 'He was on our cruiser and our cruiser is Soviet territory, regardless of whether she is in British water or not. Your suggestion that he might have been a spy is stupid. I am quite sure he was a political refugee who wanted to come to the Soviet Union, and he somehow got hurt.'

'Is that what you were told?'

'We were not told anything. The whole thing was kept secret so that the British didn't hear about it and perhaps come to take the man back and throw him into a British jail.'

When this story was published the usual controversy ensued but the general opinion appeared to be that the seamen's description clearly established the mystery frogman as Commander Lionel Crabb. Their statement that the man had been taken off by a helicopter and flown 'to a hospital ashore' tallied with the report of the Danish destroyer commander, Captain Blanders, and the contents of decoded radio messages to and from the *Ordzhonikidze*.

Once more Moscow was faced with a problem. The renewed interest in the Crabb affair had to be killed; yet a repetition of the official statement about the 'heroic death' of Crabb-Korablov would have no better effect than had the previous nine announcements. Finally they decided to launch the old story with a different presentation—and a faked 'eye-witness account' was released opportunely in Estonia.

'A number of Soviet naval officers and ratings who visited Tallin exposed how unscrupulously British authorities behave towards so-called friendly visitors when they intend to discover secrets about technical achievements in other countries,' the Estonian release stated. 'These Red Navy officers and seamen were in the cruiser *Ordzhonikidze* when the Soviet flagship, accompanied by two destroyers, brought Marshal Bulganin and Comrade Khrushchev on a goodwill visit to Britain on 18 April 1956.

'At 07.30 hours on 19 April 1956, seamen on lookout duty in the cruiser *Ordzhonikidze* observed a frogman floating between the Soviet vessels. The frogman, who wore a black rubber diving suit, with flippers on his feet, was seen on the surface of the water for a short time and then disappeared between the cruiser *Ordzhonikidze* and the destroyer *Smotriashchin.*

'The British frogman was clearly under orders to carry out an underwater spying mission on the cruiser *Ordzhonikidze,* the world's most advanced and revolutionary warship of her class. But his Anglo-American overlords did not take into consideration Soviet anti-espionage and underwater detection devices which can automatically set protective machinery into effective action.

'When the British frogman reached the cruiser *Ordzhonikidze* and started the underwater survey, the protective devices set the screw in motion and killed the ruthless Anglo-American spy. His body was eventually found and identified as the missing frogman. It was then established that the cruiser's screw had cut off his head and hands.'

This Moscow-engineered announcement was nothing but a repetition of the Soviet Note which had been delivered to the British Foreign Office on 4 May 1956. It was in complete contradiction to an earlier Soviet announcement which had stated: 'The British frogman was trapped by a security device which consists of a flap that can be aimed at the frogman and is capable of dragging him underwater to the hull of the vessel and keeping him there until his oxygen supply is exhausted. He is then released from the magnetic grip and his body is allowed to float away.'

The Estonian release was a failure. People throughout the Western world remembered the announcements that had been made in the past. They took the new release with a grain of salt and, comparing the story told by the Russian sailors in Egypt with the many reports that Crabb was alive and serving with the Red Navy, accepted the Cairo version as correct.

* * *

Patricia Rose wrote to Nikita Khrushchev; 'Would you please let me know what has happened to Commander Crabb? Is he alive? I am so miserable. I have been so worried, so unhappy . . .'

The letter was taken to the Soviet Embassy in London and handed to the Russian Ambassador, who promised that it would go to Mr. Khrushchev. But Pat never received a reply—either from Moscow or from the London Embassy.

In 1964 the Labour Member of Parliament for Brixton, Col. Marcus Lipton, examined thoroughly all the available evidence on the Crabb case. He came to the conclusion that Crabb might well be alive in Russia and that, equally, it might not be his body, after all, that lay buried in Milton Cemetery, Portsmouth. He attempted to raise the question in the House of Commons and pressed for a reopening of the case; but like all other Members who had tried, he met with failure. Nevertheless, he let it be known that he would continue his endeavours if evidence which could not be refuted became available.

Mr. Bernard Francis Castle Floud, Labour M.P. for Acton, was equally keen to hear the Commander Crabb case debated in the House. He, too, believed that the missing frogman was alive in Russia.

I had known Mr. Floud since the second world war, when he had served in the Ministry of Information. Following the publication of *Frogman Extraordinary,* we discussed all aspects of the case in considerable detail. He was already more than inclined to accept that the Russian *dossier* was genuine and he said that he intended to do a great deal of investigation personally to unearth the truth.

Shortly after his election to Parliament in 1964, I discussed the Crabb affair with him again. He was emphatic that the case must be reopened so that all the evidence could be considered and the mystery solved once and for all.

'I meant to tell you some time ago that the Russian *dossier* which you reproduced in your book is authentic,' he told me. 'I ascertained from a friend in M.I.6 that they got it through their own channels and that, with the exception of some trivial discrepencies which are clearly due to different translators using different words, the two texts are identical. For the time being this information is confidential, but I dare say that before long it will be public knowledge.'

I knew that Mr. Floud had been an intelligence officer for some ten years, that he still maintained contact with Secret Service friends, and consequently was not surprised by his disclosure—especially since Sir Percy Sillitoe, former director of M.I.5 had made a similar statement three years earlier. Mr. Floud denied that he had spoken with Sir Percy either about the *dossier* or about the Crabb case in general at any time.

'During the past few years I have investigated the matter very thoroughly,' he said. 'I established, for instance, that two naval officers from *H.M.S. Vernon* witnessed how Commander Crabb was intercepted and captured by Russian divers near the cruiser. Both officers submitted their eye-witness accounts, but their report has been kept secret and is probably carefully tucked away in a Whitehall safe. Unless an official inquiry is ordered by the Government the two officers cannot be subpoenaed to give evidence because the Admiralty would insist on the Official Secrets Act and would not permit them to disclose their discovery.

'I also discovered that some intercepted coded radio messages were deciphered and that they revealed that a frogman—obviously Commander Crabb—had been captured by Russian divers in Portsmouth Harbour; that he had been taken aboard the cruiser and then transported to Russia. Coupled with the official report from the Danish destroyer commander who observed a rendezvous between a helicopter and the Russian cruiser on the high seas and witnessed two people being taken aboard the helicopter and the helicopter flying in the direction of a Red base, the evidence that Commander Crabb was taken to Russia is overwhelming.'

Mr. Floud disclosed that much of the information contained in the Russian *dossier* had been confirmed by various independent sources in the United Kingdom and other Western countries. He claimed that the Admiralty and other government offices were aware that the body in the black frogman suit which had been washed up at Pilsey Island in June 1957, was a Russian 'plant'.

'I have examined very carefully the evidence on which the Chichester coroner relied for his verdict,' Mr. Floud said, 'and I have come to the conclusion that the inquest was what one could describe as "an inquest of convenience".

'Take, for instance, the first report the pathologist submitted. It states in black and white that there was no gross deformity of the feet and that identification was very difficult. Commander Crabb's ex-wife, who was taken to Bognor Regis mortuary to identify the body, confirmed the pathologist's findings when she told the police that she could *not* identify the corpse as the remains of her former husband because her husband had hammer toes, whereas there was no deformity whatsoever on the feet shown to her. Yet, at the resumed inquest, the pathologist testified that there was deformity of the toes and, to top it all, even spoke about the discovery of a scar on the knee which, he testified, was visible. If this contradictory evidence alone would be examined by a High Court judge, I dare say that the verdict would be vastly different from the

Chichester one.

'I am, of course, aware of the fact that the Chichester coroner considered the frogman suit as an important resource for identification because here he really had something to go on, as it was established that it had been sold to Commander Crabb. At that time no one knew about the Russian "plant" because the text of the secret *dossier* was not available until well over two years after the inquest. But even so, the possibility whether a body could have floated from Portsmouth Harbour to Pilsey Island was never examined. I went out of my way to investigate this possibility and, with the help of experts, established that the currents and tide could not have taken the body to the spot where it was found. I have charts to prove this, and I also have charts to show where the Russian submarines, which passed through the English Channel three days before the body was washed up, dumped their "plant".'

Mr. Floud concluded that the evidence—though irrefutable—at his disposal was still insufficient to force the Government to reopen the case; but like Col. Marcus Lipton, he said that if additional facts come to light he would fight for a debate in the House of Commons on the Commander Crabb affair.

15

Riddle Of A Skull

In May 1966, I received news from the U.S.S.R. that Commander Lev Lvovich Korablov had been transferred to the Far East Naval Command. In August another message followed, stating that on 7 July, 1966, Commander Lev Lvovich Korablov had been re-transferred to the Baltic Sea Fleet Command, where he was training divers and frogmen for special task duties. It concluded:

'Commander Lev Lvovich Korablov has agreed that his ten years contract with the Red Navy is extended until the age of his retirement, as he is willing to continue in the service. He was recommended by the Commander-in-Chief for promotion to the rank of Captain and it is expected that his promotion will be announced in the Red Navy Order of the Day on the 49th anniversary of the Great October Revolution on 7 November 1966.'

In December 1966, still another report was smuggled from Soviet Russia. It stated:

'On 2 August 1966, the Baltic Sea Fleet Command put Commander Lev Lvovich Korablov in charge of training *Volksmarine* (East German People's Navy) special-task divers and frogmen. Together with four other officers, Commander Korablov was sent "on loan from the Baltic Sea Fleet Command" who are in charge of the *Volksmarine* training, to the secret naval base of the German Democratic Republic (Communist East Germany) near Lübecker Bay. The duration of "loan duty" is unspecified but it is expected to be about twelve months.

'Commander Korablov immediately set up a "Special Task Diver and Frogman Training Command" at the closely guarded secret *Volksmarine* naval base near Boltenhagen. Due to the fact that the naval base is rather near the Western frontier, special precautions have been taken by State Security to safeguard complete secrecy.

'On a number of occasions traps were set by State Security to ascertain whether Korablov would yield to temptation and avail himself of the special opportunities of escaping to West Germany. When Korablov did not make any attempt even at ascertaining whether the seemingly easy escape route was in fact as it appeared to be, and quite openly ignored the opportunity, one of the *Volksmarine* officers, who was a secret Soviet State Security officer, approached him and pretended that he planned to escape across the frontier, suggesting that Korablov go with him. Commander Korablov told him that he was not interested in escaping and reported the approach to the Russian Chief Security officer. The *Volksmarine agent-provocateur* was recalled from Boltenhagen and Commander Korablov was classed as fully trustworthy—but continuous State Security supervision was not relaxed.

'On 7 November 1966, Commander Lev Lvovich Korablov was informed that in the Red Navy Order of the Day his promotion to the rank of captain had been announced. The 49th anniversary of the Great October Revolution was a day of festivities and Captain Lev Lvovich Korablov celebrated the noteworthy occasion in high spirits.'

* * *

News agency messages from West Germany announced on 19 January 1967, that the *Ostpreussenblatt* in Hamburg—the newspaper of East Prussian refugees in West Germany, which has unrivalled networks of informants in Communist East Germany—had received reports that Commander Crabb alias Captain Korablov had been seen and spoken to at the *Volksmarine* base near Boltenhagen. *Associated Press* quoted the *Ostpreussenblatt* as saying :

'It is now certain : Crabb is alive.

'Having trained Soviet frogmen, he swims now thirty kilometres* from the Lübecker Bay, near Boltenhagen. Five months ago a combat unit of divers and frogmen of the Soviet-Zonal "Volksmarine" was established there. It is Crabb's task to be responsible for the unit's training.'

This news made banner headlines in the world Press —but there were still some in Britain who stubbornly refused to accept this latest proof.

A week after the news agency's announcements I met the editor of the *Ostpreussenblatt.* He assured me that he had checked and re-checked the information received from his contacts in East Germany. He was satisfied that every word was true. When I told him that his story confirmed the reports which I had received earlier from my own contacts in Soviet Russia, he said :

'So you know that our information is accurate. The *Ostpreussenblatt* has—and this is an undisputed fact—the necessary underground contacts for obtaining top-secret information from Communist East Germany. The newspaper is the mouthpiece of the *Landmannschaft Ostpreussen E.V.*—the organisation of East Prussian refugees in West Germany—and has such unique information-networks in the Soviet Zone that their intelligence reports often are not only of considerable importance but sometimes positively sensational. There have been many occasions in the past when our people in East Germany have revealed important Soviet-Zone secrets long before anyone else in the West had any knowledge of them. Our networks have always proved to be fully accurate and sometimes of priceless value.'

* About twenty miles.

Then he told me:

'We expect a photograph of Commander Crabb in Boltenhagen to arrive in the Federal Republic soon. We know that a picture of Crabb has been taken. It is only a matter of getting it out of East Germany—though, as you surely know, it might take some considerable time before this is possible. If everything goes according to plan, our informant who talked with Crabb should arrive in the Federal Republic sometime in the Spring.'

Confirmation of the reports from Soviet Russia and East Germany came a month later from an independent Russian informant. His message said:

'Since August 1966, Commander Lev Lvovich Korablov is on loan from the Red Navy Baltic Sea Fleet Command to the People's Navy of the German Democratic Republic, where, at the secret base, Boltenhagen, the Red Navy is engaged in an extensive training programme of German divers, frogmen and other special-task units. Commander Lev Lvovich Korablov was promoted to the rank of captain on 7 November 1966, and his promotion was made known in the Red Navy Order of the Day.'

In February 1967, further details about the activities of Crabb-Korablov reached West Germany. At the beginning of March photographs of the hush-hush base of the *Volksmarine* at Boltenhagen were smuggled out. They were identified as genuine by West German experts and it was announced that they would be published as soon as those who had obtained them and got them to the West were safe from the Communist secret police.

* * *

Less than a week after this announcement from Hamburg, a sensational story broke in British newspapers. On 8 March 1967, the *Daily Express* published a front-page article under the headline: 'Buster Crabb: Skull

Find on Sands.'

It ran:

'The strange and grisly story of frogman "Buster" Crabb has opened again in Chichester with the finding of a skull.

'It was picked up in the sand by Mrs. Margaret Bull —

'*One mile* from where a headless body floated in ten years ago;

'*Ten miles* as the tide runs from Portsmouth naval base.

'Commander Lionel Crabb vanished in April 1956, when Khrushchev's Russian cruiser was lying in Portsmouth Harbour—and in the furore of speculation that followed it was suggested that he had been killed or taken prisoner while trying to find out the ship's underwater secrets.

'The torso of June 1957, was never officially identified. The skull found by Mrs. Bull on Thorney Island on Sunday may provide the final clues by four teeth in the upper jaw.

'Dr. Donald King, the Chichester pathologist who carried out a post-mortem on the torso, has now seen the skull.

'He said last night: "I would say it has been in the water about ten years. There is a clean cut across the skull, but I cannot say if this was done by a propellor.

' "Four teeth are still intact, and a dentist who attended Crabb could put the issue beyond doubt."

'Last Sunday Mrs. Bull, 26, was shrimping on the beach with her husband, Sergeant Jeff Bull, stationed at R.A.F. Thorney Island, and Chief Technician Garth Slade and his two children.

' "I thought it was a large sea shell at first," said Mrs. Bull. "It was half buried in the sand and we didn't see what it really was until we poked it with a stick."

'They wrapped the skull in a piece of cloth and took it to the R.A.F. guardroom.

'Chichester police were called. Last night a spokesman said: "We are still making investigations."

'Now the last find, the skull, may solve the mystery.'

Other British newspapers gave the skull story much less prominence. Some even ignored the item altogether. The mass-circulating *Daily Mirror* stated:

'Riddle of a Skull

'Police were yesterday investigating the possibility that a skull found on the shore at Chichester, Sussex, may have come from the body of frogman Commander (Buster) Crabb.

'Crabb disappeared in 1956 when a Soviet cruiser was in Portsmouth Harbour.

'It is believed that he was making "unofficial" underwater examination of the cruiser's hull when he vanished.

'Just over a year later, his headless body was found in Chichester Harbour.'

When these stories appeared in the newspapers I contacted a number of Crabb's close friends to hear their observations.

Chief Petty Officer Thorpe told me: 'I don't believe the skull can ever be identified as Commander Crabb's skull because it is not his. I believe that Commander Crabb is alive and in Russian naval service. I think it is not right to use the finding of some unidentified skull for connecting it in such a sensational way with Commander Crabb.'

Another of Crabb's friends and colleagues, whose name cannot be given because he is serving in the Royal Navy, said: 'I don't know why this particular skull is suddenly connected with Crabbie, especially as it has been found in an area where there were several drownings and a fatal air crash in the past and where quite a few skulls have been found in recent years. None of any previously found skulls was ever connected with Crabbie. It smells to me like an effort to counteract the steadily mounting evidence that Crabbie is alive

—like an effort to convince the public that he is dead and buried in Milton Cemetery, Portsmouth.'

'What do you think are the chances of a dentist identifying the skull as Crabb's?' I asked.

'I wish them luck,' he replied. 'You see, Crabbie was not one to go to a dentist, and I rather doubt that there exist *any* dental records of fillings in this teeth or any other treatment. Up till now it hasn't been disclosed whether the four teeth in the upper part of the skull contain any fillings.'

Commander J. S. Kerans, D.S.O., commented: 'We'll have to wait and see. Only a scientific identification of the teeth by Commander Crabb's dental surgeon—producing his records which may or may not contain precise details about dental fillings, and so forth—could be accepted as conclusive.'

A Chichester police spokesman said there was as yet nothing to identify the skull with anybody. It had been recovered in an area where in recent years, there had been several drownings and an air crash which involved fatalities. Police inquiries were continuing and a dental surgeon would be visiting the police station to examine the four teeth. All information the police could find would be passed on to the Chichester coroner, Mr. George Bridgman.

Next morning 9 March 1967, the *Daily Express* and all the other newspapers were obliged to tell their readers that an examination by a dental surgeon of the half-skull found on the shores of Chichester Harbour had not given Chichester police as much help as they had hoped.

Later that day the following report was carried:

'There were no fillings and no decay in the seven teeth in the upper jaw; the lower jaw is missing.

'The dental surgeon took X-rays of the teeth and the jawbone and expressed his view that the skull was that of a man age about 50.

'The possibilities of the dentist being able to help

in identification are now more than remote.'

The skull was never identified as Commander Crabb's—or anyone else's.

* * *

While this little mystery was having its brief day I was anxiously awaiting the promised photograph of Crabb-Korablov at Boltenhagen and news of the man who was supposed to have spoken with him there.

In April 1967, three photographs of the *Volksmarine* base at Boltenhagen reached me. The Editor of the *Ostpreussenblatt* informed me that they must not yet be released because the man who had taken them was not yet safe. He entrusted them to me to convince me that his people were trustworthy. He assured me that the photograph of Captain Korablov at Boltenhagen would be smuggled out of East Germany as soon as was practicable.

The availability of the Boltenhagen photographs was certainly proof that the East Prussian refugee movement had contacts at the top-secret base near Lübecker Bay. They could not have been obtained from any picture agency, legitimate Press photographer or any other source, and the West German authorities had confirmed that they were genuine.

There was nothing left but to wait and hope for further developments.

16

Clamp-Down At Boltenhagen

I did not meet Patricia Rose until the Spring of 1967. I had been endeavouring to locate her for years but was always told by newspaper colleagues whose help I had tried to enlist: 'She's abroad' or 'We don't know her address. She rings occasionally but never leaves her phone number.' So we were prevented from meeting although, as it turned out, the distance between our homes was only fifteen miles. It was, in fact, Pat who eventually traced me.

I learned the story which I have presented here during the course of many tape-recorded interviews which frequently took the form of cross-examination, stubborn questioning and request for proof of any statement made. Pat produced contemporary diary entries, notes, letters, British and foreign newspaper reports, photographs, and verified travel dates with passport entries to support her words.

When I questioned her statement that 'those photographs of myself which were published in Continental newspapers during the Summer of 1956 must have been taken from Crabbie's wallet after he had been captured by the Russians,' and suggested that the newspapers could have obtained the prints from some other source, she replied emphatically:

'But they couldn't have got them in any other way. It's quite impossible. Only three copies were made of the picture of me in the flowered dress—the one in which I was standing between my brother and sister-in-law, who both wore riding clothes. Mademoiselle Somborne, who took the snapshot, had one and I made sure that she still had it when Tony Kessler showed me that East German newspaper. My sister-in-law had another, and again I checked that this was so. I found it in the family album. Crabbie had the third print, because he liked the snapshot so much, and I told you that he had it in his wallet when we travelled to Portsmouth that Tuesday afternoon. Only two prints of the colour picture of me by the swimming-pool were made. Crabbie had one and I had the other. I still have mine, as you can see for yourself.'

'Perhaps your brother's manservant or someone else sold the negatives to some agency or newspaper?'

'That's out of the question,' Pat said. She promptly brought out both negatives. 'You see? I had these all the time among other films. No one could have "borrowed" them because they were mixed up with a bunch of old letters and papers. I had quite a job finding them when I wanted to make sure whether I still had them or whether someone had got hold of them. When the pictures appeared in those Iron Curtain newspapers, other people had the same idea as you. That was really why I searched so hard for the negatives.'

We discussed in detail the various messages which had come to Pat through strangers and I pointed out: 'Some of those communications are almost identical with the information given in the *dossier* which I published in *Frogman Extraordinary*. It is possible that, having only recently read my book, you might have got muddled up somehow?'

'No,' she said, rather angrily. 'I muddled nothing. These diary entries and notes were made at the time, as you can see. Apart from that, plenty of people knew about the messages because I told them whenever I got one. Then there's this French newspaper article here. It tells about Crabbie having been in Lefortovo Prison, about the Russian Colonel Myaskov having offered him a job in the Red Navy, and so on. Read it yourself. Look! It was published on 2 July, 1956, two days after

I had the first telephone call at my brother's villa in Biot. I've only just read your book during the past days but those callers told me some of the things which are in it years ago—at my brother's villa, on the train, at the *Grosvenor Hotel*, and on all the other occasions. In a way I am sorry that I read your book since it's made you jump to the conclusion that I have got mixed up or invented the messages. . . . Here! Look at these diary entries and notes and letters. Check with the people I mention there whether what I've told you is the absolute truth.'

I checked and re-checked. And many people—Pat's relatives and friends, acquaintances and semi-strangers—satisfied me that she had told them about those almost incredible messages at the time when they were received.

'Having now read my book, what is your opinion about the Russian *dossier*?' I asked her.

'Oh! it records *exactly* Crabbie's reactions—how he wouldn't say who he was, how he wouldn't admit who had sent him, how he wouldn't talk. The whole thing is Crabbie—his talk, his answers—everything is Crabbie, including the banging of the fist on the table. It's Crabbie.'

'Would you say there is any possibility that the *dossier* is a forgery?'

'No. Definitely not. It's exactly as Crabbie would have answered the questions. I can almost *hear* him. You know, I can imagine—when they gave him the cigarette—how he enjoyed that and how he said "Thank you," and little things like that. I can imagine him lying there for weeks, being very stubborn—very, very stubborn! No one could get Crabbie to do something if he didn't want to do it. He was the most stubborn little man in the world—and I can imagine him thinking to himself: "Well, I've had it." And that he said: "Shoot me if you wish, but you can't persuade me to betray my country" is precisely Crabbie. Definitely!'

'You told me earlier about that *Paris Match* picture

of Commander Crabb in Russian uniform. Did you see it?'

'I couldn't get it.'

'Well, have a look at this picture.' I handed her the photograph of the group on the Red Navy ship. 'Take your time. When you have examined it, tell me if you think you recognise anyone on it.'

She studied the photograph through my magnifying glass for some considerable time; then, pointing at the man on the left, affirmed:

'*That's* Crabbie! Is that the picture which was published in *Paris Match*?'

'Yes.' I handed her an enlargement of the man's head. 'This is the "blow-up" which they published almost full-page. It might help you to see the features more clearly.'

'It *is* Crabbie!' Pat exclaimed. 'Martin Solomon was so right when he said he was fully convinced. So am I.'

'*Why* are you so sure that it is your fiancé?'

She made a gesture with both hands. 'Well, it *is* Crabbie, that's all. I've known him more than ten years. I spent days and nights with him. I was with him continuously before he went missing . . . before he was taken to Russia. I've examined this picture, fraction of an inch by fraction of an inch, and the more I look at it the more I'm convinced it's Crabbie and nobody else. It's the face . . . I've seen Crabbie look like that. . . . Well—I *know* it's Crabbie. I wouldn't say so if I didn't believe it. I mean, the more I look at this picture, the more Crabbie comes to me. It is Crabbie. It's definitely Crabbie.'

I said gently: 'Some people have told me the man in the picture looks too young to be Commander Crabb.'

'Crabbie never looked his age. Besides, he has a face which looks quite different on many photographs.' She showed me some snapshots which had been taken at various times before Crabb's disappearance. 'If you look at these, you will see that almost on each he looks different. I have seen him so often looking down exactly as he does in the pictures from Russia. I'm convinced that it is definitely Crabbie.'

* * *

Commander J. P. A. Barr, R.N., and his wife, Freddie, who knew Crabb well for many years and who are close friends of Patricia Rose, also had not seen the photograph of the four men in Russian uniforms. Pat and I visited them at their Chelsea home and showed them the group picture. They examined it very closely.

Eventually Mrs. Barr pointed to the man on the left and said : 'This is Crabbie.'

Commander Barr nodded agreement.

'Why do you say it is Commander Crabb?' I asked.

'I've seen him so often,' Mrs. Barr replied. 'It is he —his eyebrows, his nose, his chin, the shape of his face, everything. I can see him now, looking down like that, with that cynical smile.'

'Quite a few people, including some who knew him well, insist that Crabb is *not* the man in this photograph,' I objected.

'I don't care what other people say,' Mrs. Barr retorted. 'I knew Crabbie for a good many years. I saw him often look down like that. I know his cynical smile. It is Crabbie in this picture. I am very positive about it.'

She turned to Patricia, who had remained silent. 'What do you say about it, Pat?'

'Oh! it's Crabbie. I've identified him.'

'I wanted to know *your* verdict, Mrs. Barr,' I said. 'That is why I didn't tell you that Pat and many others very close to the commander had identified him already.'

'I see. Well, I told you. I am positive it is Crabbie.'

I wanted to verify Pat's statement that she had seen Commander Barr and his wife at their home on the afternoon of Thursday, 19 April, 1956, and told them about Crabb's disappearance.

'Oh, yes, Pat came,' Commander Barr said. 'She was terribly distressed. Freddie—my wife—and I did all we could to comfort her.'

'She sobbed as she told us that Crabbie was missing. She believed we should never see him again,' Mrs. Barr added.

'Can you remember what Pat told you that Thursday afternoon?'

'Well, she said: "Crabbie's missing",' Mrs. Barr recalled, 'and I told her that she was probably worrying unnecessarily because Crabbie might only be delayed somewhere.'

'Then she told us what Crabbie had been up to at Portsmouth,' Commander Barr said. 'She told us how she had found out on the train journey to Portsmouth that he was going to "look at the bottom of the Russian cruiser." She said the Russians must have caught him because, otherwise, he would have been back by that time or at least would have phoned her.'

'Are you sure that this was on Thursday, 19 April—ten days before the Admiralty announced that Commander Crabb must be presumed dead?' I asked.

'I am positive.'

'You see, we both remember that day particularly well because we had expected Pat and Crabbie that evening. We had arranged on the previous Saturday evening, when Pat and Crabbie had dropped in, that we would have a little party here,' Mrs. Barr explained. 'But instead, Pat arrived in the late afternoon, alone—in a shocking state of nerves—and told us about Crabbie's disappearance. We did our utmost to convince her that he would come back.'

'So you didn't believe that the Russians had caught him?'

She said: 'I thought that they might have captured him and kept him prisoner on their cruiser while they were at Portsmouth, but I thought that they would set him free just before they returned home. That's why I tried to convince Pat that there was nothing to worry about. I was so confident that Crabbie would return.'

'And what was your reaction when the Admiralty announced ten days later that Commander Crabb must be presumed dead?'

'We didn't believe it,' Commander Barr said. 'Crabbie was too clever a frogman to be killed. If there had been anything wrong with his diving gear or oxygen supply he would have surfaced somehow. He did exactly that on other occasions when everything seemed hopeless. Remember, he was not just *a* frogman. He was *the* frogman—who knew *all* the answers to any tricky situation.'

'What did you think had happened when he did not return after the Soviet flotilla had left Portsmouth?'

'I had two theories,' Commander Barr replied. 'One was that Crabbie had been hidden somewhere until the storm blew over by the people who had ordered him to carry out the underwater survey. The other was that the Russians had captured him and taken him to their country.'

'Did you consider that Commander Crabb was dead, after all, when it was announced fourteen months later that a headless and handless body in a black frogman-suit had been washed up at Pilsey Island and that it was assumed to be Commander Crabb's remains?'

'We didn't believe it,' Mrs. Barr said. 'We were quite sure that it was not Crabbie's body.'

'I followed the inquest very closely,' Commander Barr added, 'and I remained unconvinced that it was Crabbie's body. The coroner clearly stated that a chain of circumstances was responsible for his verdict. The main thing that induced him to identify the torso as Crabbie's was the frogman suit which experts had recognised as one sold to him.'

'The suit was Crabb's,' I said. 'We know from the Soviet *dossier* that it had been used for "planting" the body.'

'I wonder whether the coroner's verdict would have been the same if the body had been found without Crabbie's suit and underclothing,' Pat remarked thoughtfully. 'If that had been the case I'm not sure he wouldn't have returned a verdict of "person unknown".'

'God only knows,' Commander Barr said.

'Well, there was no deformity of the big toes on the body, according to the first pathologist's report and according to the statement Crabbie's former wife made to the police,' Pat went on. 'But what does it matter what the verdict was? I and many other people know that it's not Crabbie lying buried in Milton Cemetary. He's well and alive—and I'll see him again. There's sufficient evidence to prove that. And as I see it, that's all that really matters.'

*   *   *

There were still repeated assurances from Germany that the photograph of Commander Crabb, or Korablov, at the *Volksmarine* base at Boltenhagen would arrive, but as time went by I feared that something had gone wrong.

Then, in June, 1967, a report from my informants in the U.S.S.R. arrived. It stated:

'On 19 January, 1967, as a result of international news agency announcements that Commander Lionel Crabb has been seen and spoken to at the *Volksmarine* base, Boltenhagen, Khimky Naval Intelligence Station sent a signal to the Soviet Chief Security Officer at the German naval base, requesting him to make immediate arrangements for Captain Lev Lvovich Korablov to return to the U.S.S.R. The signal also ordered the Chief Security Officer to take stringent precautions that no one contacted Korablov and to investigate who could have spoken with him and afterwards informed Western newspapermen. Special investigators left Moscow by air to supervise the security probe and support their German State Security colleagues at Boltenhagen.

'On 20 January Captain Lev Lvovich Korablov arrived by air in Moscow and was immediately driven to Khimky Naval Intelligence Station, where Colonel Yuriy Mikhailovich Antropov questioned him. Captain Korablov did not know who the Western spy was who had spoken with him and afterwards passed on the information to the West, and insisted that since his arrival in the U.S.S.R. he had never spoken about his former identity to anyone. Colonel Antropov was outraged that Captain Korablov had disregarded the strict orders from Khimky Naval Intelligence Station not to be photographed in uniform or otherwise, but he had to accept Captain Korablov's statement that he was unaware of anyone having photographed him at any time, and that if a picture of him had been taken it had been without his knowledge or permission and must have been done secretly.

'Captain Korablov was temporarly suspended from active service and sent to the State Security Committee's *dacha* near Kuntsevo, outside Moscow, where he was under continuous surveillance, so that no one could contact him. To find out whether he was trustworthy or whether he might have turned traitor, his records were carefully studied. The security officers' conduct sheets divulged that Captain Lev Lvovich Korablov was entirely uncorrupt and that he always immediately reported any suspicious approach from anyone. It was then decided to accept his explanation that he was not involved in any way with the approach from the Western spy and to redraft him to active service under improved security measures.

'During his stay at the State Security Committee's *dacha,* Captain Korablov contracted influenza which developed into dangerous pneumonia. Instead of returning to Baltic Sea Fleet Command, he was rushed to hospital, where the doctors diagnosed also another grave virus infection. His life was in the balance for almost a week but the doctors and nurses pulled him through. Recovery was slow. On 7 March, when he had regained sufficient strength for travelling, he was flown to Sochi to recuperate at the State Security Committee's sanatorium.

'Two months after Captain Korablov's arrival at Sochi the medical officer decided that he had regained his strength and was fit for active service. But, due to the recent serious illness and age, he recommended that Captain Korablov should serve with a naval command in the South to benefit from the mild climate and that he should not carry out strenuous duties. Taking his exemplary service and conduct record into consideration, the Red Navy accepted the medical officer's recommendation and transferred Captain Korablov to the Black Sea Fleet Command, Sevastopol. On 19 May, Captain Korablov reported for duty at Sevastopol and became Commanding Officer of the Special Task Underwater Operational Command.'

* * *

The photograph of Commander Lionel Crabb, alias Korablov, taken at the *Volksmarine* base, Boltenhagen, arrived unexpectedly from Germany on the evening of Monday, 24 July, 1967. When I looked at the print I was somewhat disappointed because part of the face had been torn away. The photograph was well-focused and of good quality, and there was enough left of the face for identification, but naturally I should have preferred an undamaged picture. The explanation that the photograph had been taken hurriedly and in great secrecy with a *Polaroid* camera, and that the damage had occurred when the print had been torn out too hastily, did not lessen my disappointment.

Having studied the photograph very thoroughly, I was quite satisfied that the figure in the foreground of the picture was Crabb. Nevertheless, I wanted a second expert opinion.

I contacted Patricia Rose and arranged to meet her the next day. I did not tell her about the photograph. I wanted to take her by surprise. When we met I deliberately created the impression that I was not sure what the photograph was supposed to show. I handed the print to her and said: 'Look at this and tell me if you think you know any of the men in it.'

After a single glance, she said: 'That figure in the foreground looks like Crabbie's.'

She examined the face closely, covering parts of it and looking at the print from different angles. Finally she said:

'Yes, it is Crabbie.'

'How can you be so sure?' I demanded. 'It's only half a head. It could be anybody.'

'Well, look,' she said. 'If you cover it up like this, it is very clear. It *is* Crabbie, I am sure. Look at the eyebrow . . . the nose . . . all the features. It is Crabbie. I know him with a moustache like that. I can see him sitting there now . . . lighting his cigarette and drinking his gin. I mean, I spent so many years with him. . . . Is this the picture which you expected to arrive from Germany?'

'Yes. It arrived last night.'

'Why didn't you tell me?' she asked, reproachfully. 'Why did you pretend you hadn't a clue what the picture was supposed to show?'

'I didn't want to influence you either way,' I explained.

She laughed. 'You couldn't influence me, however hard you tried. If I were not certain that was Crabbie I wouldn't say it was he.' She studied the picture again. 'Yes, I'm positive it's Crabbie. It's uncanny. Not only the face but the figure. It's Crabbie for sure.'

On Saturday, 29 July, I arranged to meet Chief Petty Officer Thorpe at Portsmouth. I used the same tactics as I had done with Pat when I handed him the photograph.

He glanced at the print and asked: 'Where was this taken?'

'I'm not sure,' I said evasively.

Like Patricia, Thorpe looked at the photograph closely, covering parts of the fact to gain a clearer impression. At last he said: 'This is Commander Crabb.'

'Are you sure?'

'Well, the eye is Commander Crabb's. So is the nose. I believe the man is Commander Crabb. These are his features, and so is the figure.'

I told him then how the picture had been obtained and added that Patricia Rose had also identified the figure in the foreground.

'That's it,' he said. 'I agree. I consider this man is Commander Crabb.'

When the story of the arrival of the new photograph broke on the radio and in the Press, Commander Crabb's cousin, Mrs. Eileen Handley, got in touch with me.

'I know Lionel is alive,' she said.

'How do you know?' I asked.

'During the past years so much evidence about Lionel being alive in Russia has come to light that I know he did not die in Portsmouth Harbour,' she said. 'Quite apart from all the messages which seem to have come from Russia and other places, the bottle-message in his own handwriting was the best possible proof.'

'I know about that bottle message,' I said, 'but some experts say that it could have been a clever forgery, and—'

'That's utter nonsense,' Mrs. Handley interrupted. 'Lionel's mother was convinced that it was his handwriting. So was I—and I still am. Why should anyone take the trouble to make such a perfect forgery, especially with a message of that sort? If it had given details about some important matters one might be inclined to entertain the possibility—but an unsigned, harmless note would not be forged by anyone.'

We talked for a long time and she corroborated many of the things which Patricia Rose and other intimate friends of Crabb had told me. Further, having been closely connected with people in government offices and having obtained firsthand information from them, she was able to cofirm several pieces of information which I had obtained confidentially and which had never been made public.

During our conversation it transpired that Mrs. Handley had seen neither the photograph of the Russian group nor the latest one from East Germany. I showed her the group photograph, without saying where or when the picture had been taken or who the men were supposed to be. I asked her to examine it and tell me whether she recognised any of the figures. She looked at the print for several minutes and then, pointing to the man on the left, said:

'This is Lionel. It looks very much like him. He was very much like my brother.' She looked at the picture again. 'Yes. It *is* Lionel . . . his eyebrows, nose, chin the whole face. Where was this taken?'

'In Russia, in the Autumn of 1959—about three and a half years after his disappearance,' I said.

'He looks quite contented,' she commented; and then: 'Oh, yes, it is Lionel.'

I showed her the East German photograph. Again she examined it carefully before pointing to the man in the foreground and saying:

'It is awfully like Lionel . . . his eye, the nose, the moustache, the whole figure. . . . I am quite sure it is Lionel.'

Mrs. Handley's mother, Mrs. Catherine Jarvis, was the next to be shown the two photographs. Her verdict, after thorough inspection of both, was similar to that of her daughter. She was absolutely certain that the pictures were of her nephew, Commander Lionel Crabb. She was particularly interested in the other photograph from Russia which shows First-Lieutenant Lev Lvovich Korablov and another Red Navy officer aboard a Soviet warship. She said with great emphasis:

'I am fully convinced that the man on the right is my nephew Lionel. It is a Godsend that his hand is so clearly visible. It excludes any possibility of mistake. It is most definitely Lionel.'

Mr. Len Cole, the landlord of the *Nag's Head* in Kinnerton Street, Knightsbridge, had known Commander Crabb since the mid-thirties and was perhaps his closest male friend. It was he who first introduced Crabb to Patricia Rose and he had later arranged their engagement party at the *Cavendish Hotel* in Mayfair. It was also he who had advised Pat to accompany Crabb on his train journey to Portsmouth on Tuesday, 17 April 1956.

'I remember that particular Tuesday very clearly,' he told me when I visited him in his charming little tavern. 'Pat and Crabbie came in about one-ish. They came daily and in a way my place was like a second home to them. Anyway, that Tuesday Pat was bubbling, as usual, but Crabbie was not his real self. He was somewhat morbid, talked very little, and I knew at once that something must be worrying him. He didn't even want his customary gin, and when I tried to find out what the matter was, he just made some vague remarks. When they left after lunch—which Crabbie hardly touched—and Crabbie asked Pat to come with him to Portsmouth on the train, she almost went into a tantrum, but I managed to make her change her mind.'

Mr. Cole confirmed in every detail Pat's account of their visit to the *Nag's Head* that Tuesday. He recalled, also, that two days later Pat had come alone to the tavern and told him Crabb was missing.

'She was in a shocking state of nerves,' he said. 'She could hardly control her tears. I gave her a gin but I don't think she knew she was drinking it. I haven't ever seen her so distressed. Anyway, when she told me about what Crabbie had been up to at Portsmouth and said that we wouldn't ever see him again, I did my best to comfort her. I pretended I was sure that Crabbie was only delayed and that we would see him in due course—but, you know, I didn't really believe that we would.'

We talked for a long time and Mr. Cole recalled the many occasions when Pat had gone to see him and told him about the messages she was receiving from mysterious people about her fiancé.

'You know, the more of those messages Pat repeated to me during the years, the more I was convinced that Crabbie was alive in Russia,' he said.

I showed him the first photograph from the U.S.S.R., picturing the four uniformed men, and asked him if he recognised anyone he might know.

'Where was this taken? And when was it taken?' he asked.

'In Russia, in the Autumn of 1959,' I replied.

He looked at the four men in the picture and after a while said, pointing to the man on the left:

'This is Crabbie.' He paused, then repeated: 'Yes, it's Crabbie. There's no doubt about it.'

When I tried to shake him by suggesting it might be merely somebody who looked like the commander, he said:

'Look, I've known Crabbie for almost thirty years. We were close friends and I know exactly what he looks like. I am positive it is him.'

I handed him the second Russian picture, showing the two men in uniform. Again he said:

'Well, it's Crabbie. It looks to me that this must have been taken at the same time as the first one because he looks down in exactly the same way. But this picture is even more evidential than the first because it shows his hand—and it *is* his hand, and that's a fact. I'm perfectly certain that it's Crabbie in both photographs.'

After he had carefully studied the photograph of Captain Lev Lvovich Korablov outside the *Volksmarine* base at Boltenhagen his verdict was unchanged: 'It is Crabbie.'

Mr. Sydney James Knowles, B.E.M., Crabb's fellow-officer and very close friend, also studied the photographs taken aboard the Red Navy vessel and at the *Volksmarine* base in East Germany most thoroughly for a very considerable time. His verdict was: 'I have no compunction in stating that the photograph of the man wearing uniform and a fur hat and also the photograph of the foreground figure in the group of three on the ice are those of Commander Lionel Crabb.'

A B.B.C. interview with Patricia Rose and myself about the latest developments in the Crabb affair, during which the technicians succeeded admirably in showing the head of the frogman on each photograph sharply and clearly over the full width of the screen, prompted more people who had known him well to get in touch with me. All testified that they had recognised him in the photographs from Russia and East Germany and insisted that there was no possibility of mistaken identity. As the reasons they gave were more or less identical with those given by all the others, it would be needlessly repetitive to publish their conclusions here.

* * *

In September 1967 my contacts smuggled another report out of Russia. This confirmed that Captain Lev Lvovich Korablov had taken up his duties as commanding officer of the Special Task Underwater Operational Command at the Black Sea Fleet Command, Sevastopol, on 19 May 1967 and stated that on 29 July he was still carrying out those duties.

On 6 October, 1967, the East German underground movement succeeded in getting to West Germany an important report which revealed:

'Since the Soviet State Security investigators arrived by air from Moscow at the *Volksmarine* base near

Boltenhagen in the early hours of 20 January, a most thorough probe has been carried out by Soviet and German State Security inquisitors. Every officer and man at Boltenhagen, and everyone who had in some way or other any sort of contact with the base, was subjected to third-degree interrogation. However, the non-stop investigation did not produce the slightest clue to who could have spoken with Captain Korablov and could have informed Western contacts. The thirty-five members of the frogman unit, as well as all the other officers and men at the base, were eventually cleared of suspicion.

'A complete security clamp-down on anything was imposed so that nothing that was going on at Boltenhagen, and especially about the training of special task divers and frogmen by Soviet experts, could leak out. The security clamp-down was so complete that not even the internal armed forces newspaper *Volksarmee* was permitted to publish any information whatsoever on the frogman theme, and consequently *Volksarmee*, which used to publish quite a lot about divers and frogmen in the past, ceased to give any information on the subject as and from the time the leak at Boltenhagen occurred.

'The investigation into the leak, which continued unceasingly, became, as time went on and no headway was made, a routine matter. Unexpectedly, the investigators were provided with an indirect lead to the identity of the sought-for informer when the photograph picturing Captain Korablov at Boltenhagen was published in the West, because the caption disclosed that it had been taken with a *Polaroid* camera. They tracked down all the people who were know to use *Polaroid* cameras and, on 10 August, arrested those who could have contacted Captain Korablov. During the interrogations the suspects were narrowed down to two and both were taken for third-degree grilling to State Security headquarters at Normannenstrasse, Berlin-Pankow.

'Despite the fact that both men could have spoken with Captain Korablov, could have taken the photograph and could have smuggled it out of the German Democratic Republic, the Soviet and German State Security investigators have arrested the wrong men. Neither of them has in fact been near Captain Korablov. Nonetheless, the interrogation is being conducted by the chief of counter-intelligence, Markus Wolf, and the Minister of State Security, Erich Mielke, is being kept informed about every single stage of the investigation.

'At present there is no evidence that either of the prisoners has been unable to withstand the brutal State Security third-degree methods and under ultra-pressure has "confessed" to having any connection with the underground movement. However, to prevent any disruption in the underground movement in case either of the arrested is broken down and talks, all communication channels have been reorganised and new links established.

'Although there is nothing yet to support this theory, it must nevertheless be anticipated that one or both men (whether guilty or not of having spoken with Captain Korablov at Boltenhagen and of having photographed him and passed everything on to the West) will be charged with espionage and sentenced by a secret tribunal. But it is unlikely that if and when such sentence is passed it would be made public because, were this done, it would establish that a leak had indeed occurred at Boltenhagen. The German Democratic Republic has repeatedly denied that there was any leak, etc., and they cannot now indirectly officially confirm by sentencing the "culprit" that Captain Korablov has been seen and spoken to at Boltenhagen and that the photograph of him is genuine.'

17

Naval Fable

The continuing attention which the international Press gave to the Crabb affair, and the mounting evidence that the frogman was alive and serving in the Russian Navy, embarrassed not only Moscow but London.

To counteract the increasing belief in Britain and many other countries that Commander Crabb was still living, somebody in British naval circles attempted to 'prove' during the Summer of 1967 that he had indeed perished in Stokes Bay on Thursday, 19 April 1956.

This new 'proof', which sounds like a James Bond fable, attempted to establish Buster Crabb did *not* dive near the *Ordzhonikidze* on the morning of 19 April, but that the frogmen spotted near the cruiser by Russian lookouts was someone else.

The 'revelation' was made—almost eleven and a half years after Commander Crabb's disappearance—by a highranking Royal Navy officer when he talked with a colleague from *H.M.S. Vernon* in a Portsmouth bar.

'The facts of the Buster Crabb case are quite different from the stories the Press and radio have put out', he said. 'Crabb was under Admiralty orders to carry out specialised underwater experiments with a revolutionary type of mine which was then on the secret list. There you have the explanation why the operation was cloaked in complete secrecy and why no one—not even the Prime Minister, Sir Anthony Eden—could give any details about it. Though this particular mine is no longer on the secret list the whole truth still cannot be told.

'Because of the rather tricky and dangerous nature of

the experiment, the location had to be outside Portsmouth Harbour and Stokes Bay was chosen as the most suitable place. You will now understand why the Admiralty announced officially on 29 April 1956, that Crabb was presumed to be dead as a result of trials with certain underwater aparatus, and why it was added that the location was in Stokes Bay.

'There was another reason why Stokes Bay was chosen. The *Ordzhonikidze* and the accompanying Soviet destroyers were at Portsmouth and if their look-outs had spotted any frogman activity in the harbour they might have concluded that we were up to some mischief. The ships had brought Marshal Bulganin and Mr. Khrushchev on a goodwill visit to Britain and everything had to be done to create an atmosphere of benevolence because our government at that time was doing its best to improve the rather strained Anglo-Soviet relations.

'Anyway, Crabb carried out the experiment with the mine. It went wrong and the damned thing exploded, but because the explosion occurred under water the detonation was muffled and audible only in the immediate vicinity of the operation. Crabb was killed and his head and hands were blown off by the blast. His body could not be found but, as we all know, it was washed up fourteen months later at Pilsey Island.'

About the other frogman, the officer said: 'He certainly did not come from *H.M.S. Vernon* or any other naval establishment. He was a freelance who had no connection whatsoever with the Admiralty or any government office and obviously was carrying out his spying for some private employer. I only know that he was not caught by the Russians and that he managed to get away to safety. Whether he succeeded in inspecting the bottom of the cruiser I couldn't say. My personal opinion is that his plan misfired because it would appear that he was spotted by the Russians before he had the chance to do what he had set out to do and had to make a run for it to save his skin.'

This story is as clumsy and unconvincing as the many Soviet attempts to make the world believe in Crabb's death. It was, in fact, what servicemen would call ruddy nonsense—but this did not deter some Continental newspaper editors from using it as 'proof' that Crabb died in 1956.

If the story is sifted rationally, and even if all the conclusive evidence that Crabb-Korablov is alive are disregarded, the following absurdities emerge:

1. If Commander Crabb were indeed on active naval service in Stokes Bay on 19 April 1956, why did he have to smuggle his frogman suit from *H.M.S. Vernon* some thirty-six hours before his alleged experiment with the hush-hush mine?* If he had been on Admiralty business, even if it were top-secret, his diving gear and oxygen supply would have been issued to him openly. Furthermore, he would have embarked on his mission from *H.M.S. Vernon* or another establishment and would have been taken to the starting point in some naval craft. Stokes Bay is more than three miles from Portsmouth Harbour—an almost impossible distance for a frogman to swim before engaging in a tricky and possibly lengthy underwater experiment. At any rate he would not have entered the water at the point near Gosport ferry where he was observed by two officers on the morning of his ill-starred exploit.

2. If, as the high-ranking officer stated, he were killed when the mine unexpectedly exploded, why did the official Admiralty announcement state: 'He is *presumed* to be dead.?' And why did ten days elapse before the official announcement was made?

3. If Crabb indeed was carrying out a top-secret underwater experiment with a new type of mine, why did Sir Anthony Eden tell a thoroughly aroused House

* It is an established fact that the suit was smuggled out of the store, together with oxygen supplies. Independent eye-witness statements prove it.

of Commons: 'It would not be in the public interest to disclose the circumstances in which Commander Crabb is presumed to have met his death.'? And why did he add: 'I think it is necessary, in the special circumstances of this case, to make it clear that what was done was done without the authority or knowledge of Her Majesty's ministers. Appropriate disciplinary steps are being taken.'? Since when are secret experiments with new weapons dependent on special authority from Her Majesty's ministers? Why should 'appropriate disciplinary steps' be taken if an experiment ended in a fatality—which is a calculated risk with most tests of new explosives? Even if in the public interest and for security reasons Sir Anthony could not disclose that tests of a very special secret weapon had taken place, he should have been able to say more than he did and should not have exposed himself to unnecessary sharp attacks from his own side of the House as well as the Opposition.

4. If Crabb were in Stokes Bay carrying out Royal Navy orders some three miles from the berth at which the Soviet cruiser lay at anchor, why did Sir Anthony tell the House during the resumed debate—with great vehemence—that all he cared for was that the outcome of the discussions with the Soviet leaders should, in truth, prove to be the beginning of the beginning and add: 'I intend to safeguard that possibility at all costs.'? Routine naval experiments miles away from the Soviet flotilla could hardly have impaired talks with the Russian leaders in any way.

5. If Crabb were indeed engaged in a top-secret underwater experiment with a mine and died because something went wrong, why was no attempt made to recover his body at the time of the accident or later?

6. If Crabb were indeed killed by the explosion of a secret mine and his body had disappeared, how could *any* naval officer state with authority that his head and both hands had been blown off?

7. If Crabb perished in Stokes Bay on 19 April 1956, how can it be explained why it took one year and seven weeks for his body to float the short distance of twelve miles, to be found at Pilsey Island on 9 June 1957—three days after Soviet submarines had passed along the English coast?

8. If Crabb were indeed killed as a result of an explosion which blew off his head and hands, why were neither the pathologist nor any other experts able to discover traces of an explosion when they examined most thoroughly the well-preserved frogman suit?

9. If the frogman who was spotted by Russian look-outs near the *Ordzhonikidze* were indeed a freelance who had attempted to survey the hull of the warship for some private employer and had succeeded in reaching safety, how can it be explained that neither the authorities, newspapermen nor private investigators ever traced him or at least discovered some clue to his identity? People who had only remote connections with the Crabb mystery were traced or came forward of their own accord. It is inconceivable that this fictitious frogman suddenly created by a high-ranking Royal Navy officer eleven years after Commander Crabb's disappearance—conveniently at a time when new evidence about Crabb-Korablov had created widespread interest throughout the Western world—should just have vanished into thin air.

10. If the frogman had indeed succeeded in escaping and if nobody were taken aboard the *Ordzhonikidze* in Portsmouth Harbour, how can it be explained that intercepted coded radio messages revealed the existence of a prisoner? And what is the explanation of the report by the Danish destroyer commander who watched the rendezvous in the Skagerrak between the *Ordzhonikidze* and a helicopter?

One of the two naval officers who had watched through binoculars Crabb's entry into the water near Gosport ferry and his interception by Russian divers—

and who had immediately submitted a written, signed report on the occurrence to his superiors—commented: 'I know nothing about any underwater experiment at Stokes Bay but that doesn't necessarily mean that there was not one which might have gone wrong. But it is nonsense to say that the frogman seen near the Russian cruiser was not Crabbie. I've known Crabbie very well for a good many years and I saw him clearly that Thursday morning. I can swear to it that it was Crabbie.'

When I acquainted Patricia Rose with the 'other frogman' story, she exclaimed:

'That's nonsense! Crabbie told me on the train journey to Portsmouth what he was going to do, and he wouldn't have lied to me. He always spoke the truth.'

It is interesting that no British or any other newspaper of repute in the Western world published the naval officer's tale. It is plain that every responsibile editor considered it too vulnerable to attack and decided to give it a miss.

Epilogue

Regardless of whether conclusive evidence about Commander Lionel Crabb being alive in Russia is produced or not, there are and always will be, members of our community who just will not accept it, but will prefer to believe that the legendary frogman is dead. And, in order to substantiate their idea, many arguments in favour of their notion are frequently brought up.

'If the Russians captured "Buster" Crabb in Portsmouth Harbour, why haven't they produced him as they did Burgess, Maclean, Philby and many others?' was a repeatedly asked question. 'Why did they announce on several occasions that "Buster" Crabb was killed?'

The answer is quite simple—the Russians could not, and cannot, admit that Commander Crabb was intercepted and captured by their divers in Portsmouth Harbour and imprisoned in their cruiser *Ordzhonikidze* because they cannot confirm that they kidnapped a British subject in British territorial waters. This was, in fact, hinted by a Soviet Assistant Naval Attaché in London on 4 May 1956, when he spoke with Mr. Gordon Holman of the *Evening Standard* and told him that 'absolutely no action' had been taken against the frogman spotted by Russian lookouts on the warship. 'We were in a British port and there was nothing we could do,' the Assistant Naval Attaché emphasized. And, to avoid any possible accusation of having infringed international law by kidnapping a British citizen in a British port, the most convenient step for the Soviet authorities was to follow the Admiralty's example and announce that Commander Crabb had died in Portsmouth Harbour on the morning of 19 April 1956.

There is another explanation why the Russians have not produced Commander Crabb publicly at a 'Press Conference' in Moscow. Burgess, Maclean, Philby, the U.S. National Security Agency officers Martin and Mitchell, and others were *defectors* where as Commander Crabb had been kidnapped in Britain and forced to join the Red Navy.

Considering the Burgess-Maclean case, which is looked upon by many as a classic example, it is interesting to recall the following facts:

Once the two defectors reached the Soviet Union in 1951 and were given political asylum as refugees, the inevitable cloak-and-dagger methods were applied—Burgess and Maclean simply disappeared. It was guessed where they had gone, but there was no proof of this and the Russians would neither admit their arrival, nor the arrival of Mrs. Melinda Maclean who with her children disappeared from Switzerland and reached Moscow in 1953.

The most remarkable feat is that Nikita Khrushchev officially declared that he had never heard of Burgess and Maclean, and later emphatically denied that they were in the Soviet Union. But in 1956, just before his and Marshal Bulganin's goodwill visit to Britain, his advisers warned him that he would be asked endless questions on Burgess and Maclean. To prevent any embarrassment to the Russian leaders by British or other newspapermen, the Soviet authorities arranged a 'Press Conference' at the National Hotel in Moscow where Burgess and Maclean met four invited journalists at a private room. During their brief confrontation with the journalists, the two defectors handed them prepared statements, assured them that they were well and happy, and then promptly disappeared.

Mr. Khrushchev's official declaration that he had never heard of Burgess and Maclean, and his denial that they were in the Soviet Union, bears some striking similarity with the statement of March 1958 in which he revealed that 'the English frogman spying on the Soviet warship has been rendered harmless' at Portsmouth Harbour on 19 April 1956. Remembering the Burgess-Maclean manoeuvre, many Western observers predicted that Mr. Khrushchev would in due course produce Commander Crabb, but the Soviet leader disappointed them.

Another attempt to discredit any evidence from Russia about Commander Crabb being alive was made by an interviewer who commented on the photograph picturing First-Lieutenant Lev Lvovich Korablov aboard a Red Navy vessel. 'Looking at this photograph which was allegedly smuggled out from Russia, I wouldn't recognise anyone if I met him in the street,' the interviewer, who had never met Commander Crabb, lashed out.

Of course he would not recognise anyone, and especially not in a frame of mind which was determined to discredit the photograph, or any other conclusive evidence, at any rate. But the photograph had not been smuggled out from Russia to enable someone to recognise Crabb alias Korablov if he accidentally 'met him in the street'. The Iron Curtain contacts had risked their freedom, and possibly even their lives, when smuggling out the 'treacherous' photograph to the West to provide pictorial corroboration of other evidence that the frogman is alive in the Soviet Union and serving in the Red Navy. Only those who knew Crabb really well are qualified to judge whether the man in the photograph is or is not the person they know intimately. And they have identified Crabb on the photographs from Russia and East Germany without a shred of doubt.

'Could not the photographs have been faked?' is still another objection to the validity of evidence on Commander Crabb being alive. 'Photographs *can* be faked, you know . . .'

Naturally, photographs can be, and often are, faked, but, knowing the vast forgery resources the Communist world has, and being familiar with their tactics, I am certain that forgery can be ruled out in this case. But let me explain this aspect in greater detail.

If, for the purpose of illustrating my point, Crabb was indeed dead and the Russians decided for some mischievous reason to produce convincing photographic 'evidence' that he is alive, their make-up experts would prepare someone to look exactly like him and their specialised photographers would produce most convincing 'proof'. And if they intended to use forgeries for sustaining their ruse, they would 'leak' *several spectacular* photographs, as they have done on previous occasions when they succeeded, for a time, to discredit and 'render harmless' awkward political opponents.

The photographs depicting Commander Crabb, alias First-Lieutenant Korablov are anything but spectacular and, despite the fact that the prints were smuggled from the Soviet Union at different occasions, both photographs must nevertheless have been taken by the same cameraman at the same time because Crabb-Korablov looks at the Ack-Ack gun on each of the Russian pictures identically. As far as the photograph from East Germany is concerned, any suspicion of forgery can be ruled out straight away—quite apart from the fact that it was smuggled out from East Germany by established underground channels which are a safeguard against deliberate 'leaks' of any sort. No skilled forger would ever consider to use imperfect pictures, and especially not one with a half torn off head, because of the difficult identification possibilities.

* * *

The purpose of writing this book was not to persuade anyone to accept the evidence that Commander Crabb is alive—it is an attempt to reconstruct the complete story on and around the legendary frogman and, by presenting all the available evidence, to enable the reader to form his own conclusions.

INDEX

A

Abbas, 121
Admiralty, 10–11, 13, 18–19, 24–5, 41, 43, 46–7, 49–52. 55–6, 59, 62, 84, 90, 93, 110–11, 113, 119–20, 123–4, 151–2, 167–8, 180–2, 186
Affray, H.M. Submarine, 11
Ali, A., 120–1
Allen, R., 106
Allen, W. H., 112
Andreyevich, V., Lieut., 138
Antropov, Y. M., Col., 170
Aremashvili, Y. P., 79
Arkhangelsk, 79
Associated Press, 156
Awbery, M. P., 105

B

Baltic Sea, 132–3, 138, 140–1, 154, 157, 170
Baltiysk, 43, 79, 102, 112
Barr, F., 7, 22, 41, 166–8
Barr, J. P. A., Cdr., 7, 22, 41, 166–8
B.B.C. T.V., 177
Beechy Head, 122–3
Berlin, 127, 138, 178
Biot, 63, 73, 82, 93, 164
Black Sea, 128, 131–2, 171, 177
Blake, E., 89
Blanders, Capt., 44, 148
Blok, 144–5
Bognor Regis, 84, 93, 152
Boltenhagen, 155–7, 161–2, 169–71, 177–9
Bosham, 87
Bostock, G. W., 88
Bridgman, G. F. L., 84–6, 90, 116, 152–3, 160
Britain, 15, 18–19, 55, 70, 94, 100, 109, 117–18, 120, 149, 156, 180–1, 187
Brockerbank, Capt., 145
Brompton Oratory, 11
Bulganin, Marshal, 10, 17, 33, 79, 94, 148–9, 181, 187
Bull, J., Sgt. 158
Bull, M., 158
Burgess, G., 9, 15–16, 186–7
Burnett, Rear-Admiral, 42, 53

C

Cairo, 150
Cap Ferrat, 73
Carkeet, H. G., Capt., 120–1, 128
Castleden, D., P.C., 85
Cavendish Hotel, 175
Chelsea, 41, 166
Chiang Kai-shek, 13–14
Chichester, 1, 83–7, 90, 93, 102, 105, 116–17, 136, 152–3, 158–60
China, 13–14, 117
Chuganov, M. M., Maj.-Gen., 8
Chuikov, L. B., Capt., 78
Churchill, Sir Winston, 60
C.I.A., 18
Clarke, A. F., Capt., 59
Cole, L., 7, 21–2, 38, 175–6
Copenhageh, 76, 136
Cordeaux, Lieut.-Col., M.P., 63
Correa, C., 121
Crabb, B., 46, 90, 94–5, 108, 146
Crabb, M. E., 7, 84–6, 89, 108, 112, 118, 132, 152, 169
Crimea, 95
Cripps, L., 28–9, 32, 114

D

Daily Express, 157, 160
Daily Mail, 119
Daily Mirror, 159
Daily Sketch, 53
Daily Telegraph, 85
Dipple, R., 21, 39
Dmitriev, V. B., 42
Dover, 63
Dugdale, J., M.P., 56–7, 59

E

Eastbourne, 120, 122
East Croydon, 97
East Germany, 11, 98, 104, 132, 154–7, 161, 174, 177–9, 188–9
Eden, Sir Anthony, 11, 56–62, 92, 106, 180, 182–3
Edwards, J., 121
Egypt, 81, 117, 146, 150
Empire News, 100, 109–10, 119, 123–4, 128, 131
England, 10, 14, 26, 52, 67, 71, 76, 80, 93, 99–100, 103, 107–8, 125–6, 130, 139, 145, 156
English Channel, 1, 86, 91, 117, 119, 127, 153
Estonia, 132, 148
Evening Standard, 54, 186

F

Falmouth, 121–2
Fernyhough, M.P., 106
Fleet Air Arm, 119
Floud, B. F. C., M.P., 7, 150–3
Foreign Office, 15–16, 53, 62–3, 106, 114, 123–4, 149
France, 23, 63, 73, 94–5, 108

French, G. R., 7, 138
Frogman Extraordinary, 103–5, 107, 109–10, 114–15, 117, 119, 130, 136, 150, 163

G

Gaczyne, 19
Gaitskill, H., M.P., 57–8, 60–3
Geddes, A. W., 90
Glasgow, 91
Golovko, A., Admiral, 132, 140–1
Gosport ferry, 28, 33, 182, 184
Granada T.V., 143
Grant, J., Capt., 54
Grosvenor Hotel, 144, 164
Grove Tavern, 21, 39

H

Hamburg, 155, 157
Handley, E., 7, 173–4
Hartlepools, 113
H.M.S. *Vernon*, 30–1, 33, 54–5, 86, 114, 151, 180–2
Holman, G., 54, 186
Home, Earl of, 124
Horsham, 23
House of Commons, 15, 56, 59–63, 105, 113, 150–1, 153, 182–3
Hunt-Crowley, J., 7, 137–8
Hutton, J. B., 9, 11, 109–10, 112–13, 115–17

I

Interpol, 127
Isle of Grain, 127, 129
Italy, 88
Izvyestia, 71

J

Jarvis, C. S., 7, 174

K

Kerans, Cdr. J. S., D.S.O., 7, 9, 106, 113–14, 117, 160
Kessler, T., 73–4, 163
Khimky Naval Intelligence Station, 47, 64, 77, 79, 130–1, 139, 169
Khorneichuk, N. P., Capt. 78
Khrushchev, N. S., 9–10, 17, 33, 79, 94, 149–50, 158, 181, 187–8
King, D. P., Dr., 83–4, 87, 89, 93, 116, 152, 158
Knightsbridge, 20, 95, 175
Knowles, S. J., B.E.M., 1, 7, 10, 55, 88, 177
Korablov, L. L., 69–70, 76–9, 102, 104, 107–8, 112–13, 128, 130–5, 138–42, 148, 154–7, 161, 169–71, 175, 177–9, 182, 184, 188–9
Kotov, V. P., Rear-Admiral, 42, 53
Krasenko, K. O., 68–9
Kremlin, 14, 16, 100
Kronstadt, 78, 140, 142
Kutzov, First-Lieut., 45

L

Lamport, S. J., Det.-Supt., 43, 52
Lefortovo Prison, 52, 64, 76, 163
Leghorn, 88
Leningrad, 70, 78, 80, 132, 135, 140–1
Lipton, Col. M., M.P., 150, 153
Lloyd, Selwyn, 105–6, 113–14
London, 20, 26, 29, 38, 41, 51, 53–4, 63, 71, 79–80, 90, 95–8, 105, 123, 131–2, 144–5, 150, 180, 186
Lopes, T. J., 121
Lowndes Arms, 29
Lübecker Bay, 154, 156, 161

M

Maclean, M., 187
Maclean, R. D., 9, 15–16, 186–7
Macmillan, H., 92
Martin, W. H., 187
Mayfair, 175
McLanachan, Lieut., 86, 91
Middle East, 94, 144
Mielke, E., 178
Milton Cemetery, 1, 92–3, 102, 150, 160 169
Mitchell, B. F., 187
M.I.5, 16, 136, 151
M.I.6, 151
Monck's Bar, 35
Moscow, 1, 13–17, 19, 47, 52, 54, 64–5, 72, 74–6, 96, 101, 103–4, 124, 126, 130–1, 138, 148–50, 169–70, 177, 180, 187
Mr. X, 13, 15–17, 19, 41
Muralov, A. G., Capt., 81
Murmansk, 141
Myaskov, A. P., Col., 64–9, 71–2, 76–81, 104, 107, 111, 163

N

Nag's Head, 21–2, 38–9, 175
National Hotel, 187
Newcastle-under-Lyne, 119
News of the World, 113
Nore, C-in-C, 122
Niton Radio, 122
Nizhansky, A. B., Capt., 102, 112
No Banners, 112
N.T.S., 76

O

Odessa, 131–2
Official Secrets Act, 53, 57, 137, 151
Ordzhonikidze, 1, 11, 13, 17–18, 26, 33, 35–7, 39, 42–4, 49–55, 68, 76, 87, 91, 94, 101, 146, 148–9, 180–1, 184, 186
Ostpreussenblatt, 155–6, 161

P

Paris, 93
Paris Match, 143, 164–5
Peking, 117
Pendock, Mr., 20, 26, 40–2
Pentagon, 18
People, The, 112
Philby, H. Kim, 15–16, 186–7
Pilsey Island, 83–6, 94, 152–3, 168, 181, 184
Poland, 11
Polish Club, 145
Port Said, 119, 125–7, 129
Portsmouth, 1, 10–11, 13, 18, 20–3, 26, 28, 30, 33–4, 38–40, 42–3, 52–6, 64, 67, 73–6, 79, 82, 84, 86—7, 92–4,101–2‘ 112–15, 117, 120, 131, 136, 139, 146,

Portsmouth (*cont.*)
150, 152–3, 158–60, 163, 167–8, 172–3, 175–7, 180–2, 184–6, 188
Pravda, 71
Price, D., 39–40
Prides, L., 43
Purtsov, I. A., 102

R
R.A.F., 83, 122, 158
Randall, J., 83, 87
Rankin, M.P., 91–2
Red Cross Society, 124
Red Navy, 75–8, 81, 96, 99, 102, 107, 113–16, 119, 125, 127, 130–1, 133, 135–6, 138–41, 144–5, 148, 150, 154–5, 157, 163, 165, 168, 171–2, 175, 177, 187–8
Rennie, W., Chief Officer, 121
Reuter, 51
Reynolds News, 79
R.F.A., 119–20, 124
Richman, E., 28, 43, 114
Riviera, 73
Robinson, W., 7
Rome, 144
Rose, P., 7, 17, 20–7, 38–44, 46–7, 63, 71, 73–6, 82, 93, 95–8, 108, 117, 143–6, 150, 162–9, 172–7, 185
Royal Sovereign lightship, 122, 128
Russia, 9, 11, 16, 18, 54, 80, 96, 98, 103, 105, 116–17, 127, 130, 133, 136, 138, 140, 143, 145–6, 150, 152, 154, 156–7, 165–6, 173, 176–7, 186, 188

S
Sallyport Hotel, 28, 31, 35, 41, 52, 114
Savelyeva, N. N., 70
Scandinavia, 94, 133
Secret Service (British), 13, 60, 63, 151
Secret Service (Soviet), 14–16, 30, 126
Selskoye, 68, 77
Sevastopol, 102, 128, 141, 171, 177
Shcherkin, A. A., 42
Shinwell, E., M.P., 58
Sillitoe, Sir Percy, K.B.E., C.B.E., 7, 136, 151
Simmonds, S. L., Supt., 64
Skagerrak, 44, 136, 184
Slade, G., Chief Technician, 158
Smirnov, B. M., Major, 71, 80
Smirnov, R. V., 125, 127, 129
Smith, D., 123, 128
Smith, M., 15–18, 24–6, 28–30, 32–8, 41, 43, 49, 52, 71–2, 74, 91
Smith, W., 119–25, 128–9
Smith, W. R., Third Officer, 119–29
Smotriashchin, 36, 53, 149
Sochi, 171
Solomon, M., 143–4, 165
Somborne, Mlle, 73–4, 165
Southern T.V., 108
Special Branch, 16
Squirrel, H.M.S., 122
State Security Committee, 47, 68, 70–1, 77, 100–2, 104, 107, 131–2, 138–9, 141, 145, 155, 170–1, 177–9
Stiepanov, G. F., Capt., 101
Stoke-on-Trent, 119, 128
Stokes Bay, 43–4, 50, 52–3, 56, 180–5
Sunday Pictorial, 110
Sverdlov, 14, 18, 30, 34, 52
Switzerland, 187

T
Tallin, 148
Tass, 80
Thomas, A., 89
Thomas, J., 7, 110–12
Thorney Aerodrome, 83
Thorney Island, 158
Thorpe, A. R., B.E.M., 7, 114–16, 135, 159, 172–3
Times, The, 44, 56, 59
Trobuz, Admiral, 76
Truculent, H.M. Submarine, 11

U
United Kingdom, 86, 152
U.S.A., 18–19, 26, 145
U.S.S.R., 14–15, 36, 41, 44, 80, 87, 100, 103, 107, 123–7, 130, 133, 135, 139, 141, 144, 148, 154, 156–7, 169–70, 176, 187–9

V
Victoria, 22, 96, 144
Vladivostok, 98–9, 102, 130–1, 133, 139, 143
Volksarmee, 178
Volksmarine, 154–7, 161, 169, 171, 177–9
Vosensky, K. G., 42
Vnukovo Airport, 47
Vutnovko, I., 44

W
Wallbridge, Miss, 125
Ward, Dame Irene, M.P., 105
War Office, 15–17
Wave Commander, 119–20, 122, 124, 127–9
Whitehall, 79–80, 110, 151
Wolf, M., 178

Y
Yalta, 131
Yangtse, 9, 113
Yevdokimov, B. F., Lieut.-Col., 130

Z
Zhabotin, I. S., Col., 46–51, 64